FIELD GUIDE TO THE
ORCHIDS
of Europe *and the* Mediterranean
SECOND EDITION

FIELD GUIDE TO THE

ORCHIDS

of Europe *and the* Mediterranean
SECOND EDITION

Rolf Kühn
Henrik Æ. Pedersen
Phillip Cribb

Kew Publishing
Royal Botanic Gardens, Kew

© The Board of Trustees of the Royal Botanic Gardens, Kew, 2024
Text © Rolf Kühn, Henrik Æ. Pedersen and Phillip Cribb
Photographs © Rolf Kühn unless otherwise indicated in the captions and on p. 417.

The authors have asserted their rights to be identified as the authors of this work in accordance with the Copyright, Design and Patents Act 1988.

All rights reserved. No part of this publication may be reproduced, stored in a retrieval system, or transmitted, in any form, or by any means, electronic, mechanical, photocopying, recording or otherwise, without the written permission of the publisher unless in accordance with the provisions of the Copyright Designs and Patents Act 1988.

Great care has been taken to maintain the accuracy of the information contained in this work. However, neither the publisher, the editors nor the authors can be held responsible for any consequences arising from use of the information contained herein. The views expressed in this work are those of the individual authors and do not necessarily reflect those of the publisher or of the Board of Trustees of the Royal Botanic Gardens, Kew.

Second edition published in 2024
First published in 2019 by the Royal Botanic Gardens, Kew, Richmond, Surrey, TW9 3AB, UK. www.kew.org

ISBN 978-1-84246-819-7
e-ISBN 978-1-84246-820-3

Distributed on behalf of the Royal Botanic Gardens, Kew in North America
by the University of Chicago Press, 1427 East 60th St, Chicago, IL 60637, USA.

British Library Cataloguing in Publication Data
A catalogue record for this book is available from the British Library.

Production Management: Georgina Hills
Design and page layout: Christine Beard and Nicola Thompson
Copy-editing and proofreading: Sharon Whitehead

COVER ILLUSTRATIONS
Front: *Ophrys tenthredinifera*, Italy, Sicily
Back, left to right: *Orchis italica* Greece, Crete (25); *Neotinea lactea*, Italy, Sicily; *Ophrys cretica*, Greece, Crete; *Cypripedium calceolus*, Italy, Dolomites (25); *Serapias vomeracea* subsp. *laxiflora*, Greece, Pelion (25); *Platanthera chlorantha*, Switzerland, Zürich Oberland (25)I

Printed and bound in Italy by Printer Trento srl

For information or to purchase all Kew titles please visit shop.kew.org/kewbooksonline or email publishing@kew.org

Kew's mission is to understand and protect plants and fungi, for the wellbeing of people and the future of all life on Earth.

Kew receives approximately one third of its running costs from Government through the Department for Environment, Food and Rural Affairs (Defra). All other funding needed to support Kew's vital work comes from members, foundations, donors and commercial activities, including book sales.

CONTENTS

Foreword	vii
Preface	ix
Introduction	1
The orchid plant	2
The structure of orchid flowers	4
Pollination	6
Ecology	9
Orchid habitats	10
Conservation	22
Classification	25
Taxonomy of European orchids	27
Taxonomic concepts	27
How to use this field guide	35
Genera and species	37
Natural hybrids	390
Acknowledgements	417
Appendix 1: Keys to genera and species	419
Appendix 2: Bibliography	437
Index of scientific names	443
Index of common names	461

Anacamptis morio subsp. *longicornu*
Italy, Sardinia

FOREWORD to second edition

Orchids are famous for having forged mutually advantageous relationships with other organisms. Most obviously, they bewitch, beguile or even shamelessly seduce visiting animals into couriering their pollen masses from flower to flower, often requiring the visitor to forgo any form of payment. Less obvious, but equally important, are the vast networks of fungal hyphae that both kick-start orchid seeds into growth and provide essential nutrition to more mature plants; the fungi even plumb some orchids into adjacent trees to provide a free lunch.

But orchids have a third partnership that today is equally as important for their future – a partnership with humankind. There is of course an increasing temptation to view orchids as our victims rather than our partners. Orchid species become extinct on a weekly basis and we are rarely entirely innocent in their demise, the charges levelled against us varying from manslaughter in the case of climate change through to murder when orchid habitats are destroyed or particularly alluring species are over-collected in the wild. The myriad of man-made tropical orchid hybrids now cheaply available from supermarkets worldwide can hardly compensate for losses from the wild, yet those rather unsubtle hybrids offer useful reminders of one important fact – (almost) all of us like orchids. Familiarity may have slightly eroded their mystique but it shows no sign of breeding contempt.

No wonder then that orchids are the preferred botanical pin-ups for so many conservation organisations, or that most European countries have formed energetic societies that specialise in exploring their own native orchids. Orchids have acquired a loyal following of people every bit as diverse as the orchids they love.

Here we have a book co-authored by three experts who have, over the years, demonstrated extraordinary commitment in pursuit of orchids. Now they have pooled their collective knowledge to bring us the latest in a diverse line of European orchid floras, featuring well-chosen background information followed by a systematic treatment constructed within a pragmatic taxonomic framework. High-quality images, informed text and newly added keys to the genera and species are dovetailed into a beautifully presented, user-friendly volume that will delight many European orchid aficionados. Yet perhaps its greatest long-term contribution could prove to be catalysing further research. The more we learn about orchids, the more fascinating they become.

Richard Bateman
President of the Hardy Orchid Society and
Honorary Research Associate at the Royal Botanic Gardens, Kew

Gymnadenia conopsea, albino form
Italy, Umbria (21)

PREFACE

It is perhaps counterintuitive to say that we still have large lacunae in our knowledge of European and Mediterranean orchids despite decades, even centuries, of examining and studying them. Their association with mycorrhizal fungi remains poorly studied. We do not even understand why some fungi have mycorrhizal properties and others that are morphologically similar do not. Nor do we know how specific the relationship is between orchids and fungi. Similarly, orchid pollination, which is used for example to distinguish taxa in *Ophrys*, is imperfectly understood. Many such studies have observed the removal of pollinia from orchids by various insects, but fewer observations have ever been made of the deposition of the pollinia on the appropriate stigma of another flower.

Similar gaps exist in our understanding of orchid phylogeny and classification. We have a much-improved understanding of phylogeny at the generic level thanks to advances in DNA analysis and the increasing power of computing to analyse large data sets. However, many early studies examined a limited number of taxa and often only one sample of each. Consequently, several major changes in orchid classification have resulted from modern phylogenetic studies, a few of which are perhaps premature However, the number of studies involving DNA technologies and data analysis has increased exponentially. Nowadays, more complete sampling of taxa and analyses of greater numbers of genes underpin several recent reconsiderations of generic and species delimitation in our native orchids. In this second edition, we have attempted to take these studies into account. Thus, here we have resurrected *Coeloglossum* and *Nigritella* which previously had been subsumed, respectively, into *Dactylorhiza* and *Gymnadenia* (see Taxonomy of European Orchids). Both genera are morphologically easily distinguished in the field and we believe that this move will be widely welcomed. Other changes at the specific and infraspecific levels arising from either DNA or morphological analyses have been incorporated in *Epipactis* and, less comprehensively, in a few other genera.

As a feature in this new edition, identification keys to all the genera and species are provided (see Appendix).

New images have also been added where they help in the identification of difficult taxa. Various colleagues have provided useful comments on the first edition, enabling us to improve some of the distribution maps, and we are grateful for their assistance.

Rolf Kühn, Henrik Æ. Pedersen and Phillip Cribb
July 2024

Anacamptis papilionacea subsp. *papilionacea*
Italy, Puglia (25)

INTRODUCTION

Europe has a distinctive and rich orchid flora that extends from the Atlantic islands in the west, south to the Mediterranean fringe of North Africa and east to the Urals and the Middle East. The deserts of the Sahara and Middle East form a barrier to the south and the permanently frozen Arctic to the north. The range of a few European orchids (e.g. *Calypso bulbosa, Coeloglossum viride, Cypripedium calceolus, C. guttatum, C. macranthos, Epipactis helleborine, Epipogium aphyllum, Goodyera repens, Herminium monorchis* and *Ponerorchis cucullata*) extends east to China and Japan, while a few also reach the Himalayas. Narrow endemics, such as *Habenaria tridactylites* and *Goodyera macrophylla*, are rare and mostly confined to islands in the Canaries and Madeira, respectively.

Orchids are a distinctive family characterised by their lifestyle, which encompasses the production of microseeds that have an embryo and seed coat but lack an endosperm (food supply). The nutrients needed for germination and growth are provided at and after germination by a fungal symbiosis, termed a mycorrhiza, that penetrates the body of the embryo. The orchid embryo breaks down the fungal threads (hypha) within its cells, releasing nutrients that the orchid then utilises. A minority of orchids, including the European bird's nest (*Neottia nidus-avis*) and the ghost orchid (*Epipogium aphyllum*) rely for their entire life cycle on nutrients derived from their fungal partner. Most, however, soon develop roots and green leaves, the latter allowing them to photosynthesise and grow.

Neottia nidus-avis growing in coniferous woodland, N Italy (21)

INTRODUCTION

Orchids are one of the largest families of flowering plants with an estimated 28,000 species (Govaerts *et al*. 2023). The family has a worldwide distribution, the majority having a tropical or subtropical distribution and many growing as epiphytes on trees or lithophytes on rocks. In the tropics and subtropics, terrestrial orchids are often in a minority, albeit still well represented. By contrast, all European and Mediterranean orchids are terrestrial, growing in soil or leaf litter.

European orchids are perennials and therefore have a distinct dormancy period. In the Mediterranean, they grow during the mild wet winter season, usually flowering towards the end of this period in the spring or early summer. The plants move into dormancy as the summer develops with dry warmer weather dominating. Further north, the growing season begins in early spring and terminates with flowering in summer or early autumn and the dormancy period occurs in late autumn and over winter. Dormancy is ensured by storing food in storage organs, which allow a rapid response when suitable growing conditions recur.

Orchids often have attractive flowers, nowadays the main reason for their popularity. However, both in habit and in floral structure, they are deceptively simple. A thorough understanding of orchid morphology is essential for anyone trying to identify them.

THE ORCHID PLANT

The name orchid comes from *orchis*, a Greek word that means testicle, referring to the paired underground tubers (strictly tuberoids) that are found in many but not all European orchid species. These tubers store starch during the dormant season which is used when the orchid starts to grow again when appropriate weather resumes.

In some genera, notably *Cypripedium*, *Goodyera*, *Epipactis* and *Cephalanthera*, the storage organ is the rhizome, a branching underground stem system. A third kind of storage organ, a swollen short stem called the pseudobulb, is found in the genera *Calypso*, *Hammarbya*, *Liparis* and *Malaxis*.

European orchids have entire leaves, never lobed ones. Two types of leaves can be found in European orchids. *Cephalanthera*, *Epipactis* and *Cypripedium* have pleated leaves. The rest have unpleated (conduplicate) leaves. Leaf number can vary from one to several depending on the species. Thus, *Calypso*, *Malaxis* and *Steveniella* have a single leaf, whereas most *Anacamptis*, *Cephalanthera*, *Cypripedium*, *Epipactis*, *Ophrys*, *Orchis* and *Serapias* have several. In *Corallorhiza trifida*, *Epipogium aphyllum*, *Neottia nidus-avis*, *Limodorum abortivum* and *L. trabutianum* the leaves are reduced to sheaths and scales.

Most orchids are most easily identified from their flowers. The basic structure follows that of other monocotyledons with their parts in threes. Thus, orchids have an inferior ovary whose tripartite nature can be seen in cross-section. The ovary usually takes the function of a pedicel. In the bud stage, the lip faces upwards. The lip achieves its mature position by twisting through 180° (resupination) and serves the pollinator as a landing platform. Very seldom no rotation occurs, as in *Epipogium aphyllum*, or there is a rotation by 360°, as in *Hammarbya paludosa* and *Malaxis monophyllos*.

Orchids have a calyx comprising three sepals, with the dorsal one sometimes a different shape and colour from the lateral ones. The corolla comprises two lateral petals which can often be rather insignificant, certainly often smaller than the sepals. The third petal is very different from the lateral petals, usually being larger and more complex in structure. It is called a lip or labellum and is important in pollination, often acting as an attractant for potential pollinators. The central parts of the orchid flower are greatly modified when compared to those of other monocotyledons. The number of parts of the androecium (male) and gynoecium (female) organs are greatly reduced and these parts are fused into a single structure called the column. In most European orchids, the number of stamens is reduced to one and it is placed terminally on the column, the exception being *Cypripedium* which has two lateral stamens and a sterile terminal one called a staminode. Another peculiarity of orchids is that the pollen is not free but aggregated into pollen masses (called pollinia), which can be stalked and attached to a sticky mass (the viscidium) to enable a pollinator to remove them en masse. The stigma is placed ventrally on the column and is either tripartite or bipartite. One lobe of the stigma can be sterile and form a rostellum that separates the stigma from the stamen. A summary of the various flower types is presented below.

Franz Bauer's watercolour of *Orchis mascula*, showing the paired tubers

THE STRUCTURE OF ORCHID FLOWERS

Cypripedium

1 dorsal sepal
2 synsepal
3 petal
4 lip
5 stamen
6 staminode
7 ovary

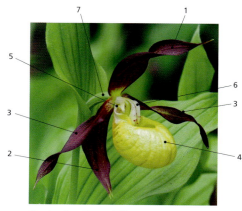

Cypripedium flower (53)

Epipactis

1 dorsal sepal
2 lateral sepal
3 petal
4 lip hypochile: rear part of lip
5 lip epichile: front part of lip
6 column and anther
7 ovary

Epipactis flower (21)

Epipogium

1 dorsal sepal
2 lateral sepal
3 petal
4 lip
5 spur
6 column
7 stamen

Epipogium flower (21)

THE STRUCTURE OF ORCHID FLOWERS | 5

Platanthera

1 dorsal sepal
2 lateral sepal
3 petal
4 lip
5 spur
6 column
7 anther loculus
8 stigma
9 ovary

Platanthera flower (21)

Orchis

1 dorsal sepal
2 lateral sepal
3 petal
4 lip
5 anther loculi
6 bursicle
7 mouth of spur
8 spur

Orchis flower (21)

Ophrys

1 dorsal sepal
2 lateral sepal
3 petal
4 column
5 lip basal field
6 lip pattern shield (speculum)
7 lip apical appendage

Ophrys flower

POLLINATION

Pollination is the transfer of pollen onto the stigma of the plant. If successful, the pollen grains germinate and send a pollen tube carrying the male gamete down the style to eventually enter an ovule so that fertilisation can occur. In European orchids, we find both autogamy (self-pollination) and allogamy (cross-pollination).

Most European orchids have flowers that have evolved to be pollinated by particular insects. Thus, the form of their flowers is intimately linked to the insects that pollinate them.

Why are insects attracted to orchid flowers? In most cases, pollinating insects are looking for food, either nectar or pollen, but in *Ophrys* they are looking for a mate. The form of the flower, its colour or its fragrance can be attractive to pollinators. Thus, *Pseudorchis albida* and *Epipactis atrorubens* flowers typically produce a scent in which the fragrance of vanilla is dominant, whereas *Herminium monorchis* smells intensively of honey. Not all orchids smell so pleasantly. Unpleasant odours emanate from *Anacamptis coriophora* (smells of bedbugs), *Orchis mascula* and *Himantoglossum hircinum*, the last smelling strongly of goats.

The fragrance of some orchids, for instance *Platanthera bifolia* and *P. chlorantha*, intensifies at dusk and attracts moths.

A number of European orchids produce no nectar but, by resembling other nectar-producing flowering plants, can be pollinated by insects that are attracted to the plant that the orchid mimics.

Insect pollinators

a) Short-tongued flies and beetles
Orchids which are pollinated by flies and beetles usually have white, yellowish, brownish or greenish flowers and smell pleasantly, for instance, *Neottia nidus-avis* and *N. ovata*. To attract these insects, it is imperative that the nectar is easily accessible. *Neottia ovata* is often pollinated by the beetle *Cantharis rustica* (Cantharidae). The pollen is transferred from the head of the beetle to the stigma of the next flower to be visited.

b) Butterflies and moths with long tongues
Orchids such as *Anacamptis pyramidalis*, *Gymnadenia conopsea* and *Gymnadenia odoratissima* have purple, pink and white flowers that are strongly scented and attract butterflies

Gymnadenia densiflora and six-spot burnet moth carrying pollinia on its tongue (25)

and day-flying moths, such as burnets. Yellow and white-flowered orchids with long nectariferous spurs, such as *Platanthera chlorantha* and *P. bifolia*, are mostly pollinated by moths that are active at dusk and night-time.

c) Bees, bumble bees and wasps

In most cases, orchids that are pollinated by bees and bumble bees, such as the species of *Orchis*, are conspicuously coloured. In some cases, bumblebees with short tongues gnaw through the floral spur to get to the nectar (if any). Because the spur is adapted to the length of the tongue of the pollinator, only a limited spectrum of insects can pollinate the flower.

The pollination of the species of *Ophrys* is amongst the most remarkable of all. Newly emerged male bees (or, in a few cases, other male insects) are attracted to *Ophrys* flowers by their fragrance, which mimics that of female bees, and by their shape and texture, which also resemble those of female bees. The bee lands on the lip of the flower and tries to mate with it, thereby removing the pollen masses that stick to his head or abdomen. If the bee repeats the process on another flower, pollen can be transferred to the stigma of the second flower because the pollen mass stalks wilt while they are attached to the bee, placing the pollen masses in a position that allows them to enter the stigmatic cavity of the new flower. *Ophrys* pollination seems to have a high degree of specificity, at least at subspecies level. For example, *Ophrys insectifera* subsp. *insectifera* is usually pollinated by the digger wasp *Argogorytes mystaceus* (less often by *A. fargeii*).

Occasionally a pollinating insect will visit the flower of a species other than that from which it took the pollen and hybridisation may occur.

Wasps (e.g. *Vespula* spp.) and, less commonly, bumble bees, pollinate *Steveniella satyrioides* and certain species of *Epipactis*.

The small wasp *Argogorytes mystaceus* on a flower of *Ophrys insectifera* (13)

The solitary bee *Eucera longicornis* on a flower of *Ophrys scolopax* (56)

The solitary bee *Andrena barbilabris* carrying pollinia from *Ophrys sphegodes* subsp. *aveyronensis*

Highly specialized pollination apparatus are found in the species of *Orchis* and *Dactylorhiza*. Their pollen masses have adhesive discs that are enclosed in a sack (bursicle), which is filled with a sticky liquid. Thus, if a bee visits a flower of *Orchis mascula*, it comes into contact with the sack, which releases the sticky disc attached to the pollen and hence attaches the pollen to the insect. The stalk of the pollen mass gradually bends forward so that, if the insect visits another flower, the pollen mass comes into contact with the stigma and pollinates the flower.

Bombus lucorum on *Epipactis atrorubens* (21)

Cypripedium, with its slipper-shaped lip, has a pollination syndrome that is unique in European orchids. Certain small bees such as *Andrena haemorrhoa* are attracted to the glossy staminode of *C. calceolus*, but bees that visit the flowers of this species frequently slip into the lip, which is effectively a trap that does not allow the bee to escape by flying out of the lip mouth. The trapped bee is then attracted by the bright translucent side windows at the back of the lip, which it approaches by passing the stigma (depositing any pollen acquired from a different flower as it does so) and then one of the stamens, so that it leaves the trap through one of the windows with new pollen mass on its head.

A small *Andrena* trapped in the lip of *Cypripedium calceolus* (25)

Self-pollination (autogamy)

Self-pollination involves the pollen masses of a flower entering the stigma of the same flower. *Ophrys apifera* is an exception in its genus because, although insects visit it, it is almost always self-pollinated. Pollination is achieved when the limp pollinium stalks fail to hold the pollen masses in an upright position so that they fall out of the anther and pivot onto the stigma. Self-pollination is also found in other genera, for instance in some species of *Epipactis* and *Cephalanthera* where the rostellum, the structure separating the stamen

Ophrys apifera showing the pollinia falling out of the anther loculi (56)

from the stigma, has been lost. Then, the pollen can drop or ooze onto the stigma, which lies below and behind the stamen, under the influence of gravity.

ECOLOGY

European orchids grow in a range of diverse habitats. They are often primary colonisers of bare soil, favouring places such as new roadside verges and cuttings, soil disturbed by grazing animals and other sites where open places devoid of other plants are frequent. Many species, but by no means all, are very particular about where they grow. The climate, soil and pH (all abiotic factors), as well as mycorrhizal associations and competition from other organisms (biotic factors), can be critical. Upsetting any of these can rapidly lead to the loss of orchids from any site. Some of these factors are considered below:

Climate

The richest region for orchids in Europe is the Mediterranean, where there is a wet winter growing season and a dry summer, during which the orchids become dormant. In central and northern Europe, the growing season runs from spring until early summer, with orchid dormancy running from late summer through the winter. Some orchids are confined to the Mediterranean region, others to the more temperate climates of north and central Europe. A number of Mediterranean species occur quite far north where the microclimate allows. A few orchids can even survive within the Arctic Circle.

Substrate

All European orchids are terrestrial, the vast majority growing in soil or occasionally in crevices in rock. Orchids are most commonly found on basic soils, such as chalk and limestone, less so on neutral and acidic ones, although a few orchids require acidic conditions. A few orchids, notably some *Dactylorhiza* species, *Epipactis palustris*, *Hammarbya paludosa*, *Liparis loeselii* and *Malaxis monophyllos* grow in moss in fens, marshes or bogs.

pH Value

pH provides a measure of the acidity or alkalinity of a substrate. The pH scale runs from 0 to 14: pH 7 is neutral, a pH below 7 is acidic, and a pH above 7 is basic (alkaline).

Many European orchids favour neutral or basic (alkaline) substrates, such as chalk or limestone. Others are not so demanding and can grow on neutral and slightly acidic soils. A few orchids, such as *Dactylorhiza maculata* subsp. *maculata*, grow on acidic soils.

Mycorrhizal fungi

The mycorrhizal fungi associated with orchids provide nutrients to the developing embryo and plant. The hyphae (fungal threads) of the fungus penetrate the orchid through the latter's root hairs. The orchid cells break down the hyphae and extract nutrients from the fungus. The specificity of the orchid–fungus relationship is variable.

As a rule-of-thumb, the more common orchids are less particular than the rarer ones. In fact, the distribution of orchids is just as much a reflection of the needs of their fungal symbionts as it is of the orchids themselves.

Competition

Most orchids are primary colonisers of bare soil, although many of them only need tiny patches of open soil (for germination) in otherwise closed vegetation. If competition with other plants becomes severe, the orchids decline, particularly if the invading species are woody. The vegetation of the best orchid sites in grassland and scrub is maintained at a sub-climax level by grazing or regular disturbance, such as that which occurs where agricultural practices are non-intensive. Some orchid species, however, are specialised forest plants and their long-term survival often depends on careful (or no) forestry practice combined with undisturbed hydrology.

ORCHID HABITATS

Orchids are found more or less throughout Europe and the Mediterranean region in both natural habitats and man-made ones. They are absent only from most urban environments, intensively farmed land, open water and the upper regions of the highest mountains.

There is a gradation in number of species in Europe from south to north and from sea level into the mountains. The Mediterranean climate and soils, where limestone is often the base rock, provide the area that is richest in terms of species number. Acidic bogs and grassland are less productive for orchids, although some are found in such habitats. A few of the richest habitats are discussed below.

The Mediterranean

The Mediterranean climate is defined by the ability of the olive (*Olea europea*) to flourish. The main habitat around the Mediterranean is known as 'garigue' in France and 'phrygana' in Greece and is particularly rich in orchids. In particular, it is the home of many species of *Ophrys* and *Serapias*. The low, spiny and fragrant scrub with frequent bare or lightly vegetated patches is favoured because competition from other plants is light. Heavy grazing is often a problem in this region, especially because goats and sheep favour the succulent flowering stems of orchids.

In the Mediterranean, the open woodland (called 'maquis' in France) that develops from 'garigue' when grazing levels are low is also a rich orchid habitat, with *Orchis* and *Anacamptis* species particularly common. 'Maquis' contains trees such as evergreen oaks, carob (*Ceratonia siliqua*) and the Judas tree (*Cercis siliquastrum*), as well as many spiny and fragrant shrubs, such as Mediterranean gorse (*Calicotome villosa*), rockroses (*Cistus* and *Helianthemum*), sage (*Salvia*) and mastic (*Pistacia*) bushes. Even in more closed woodland, the glades, rides and margins can be rich in orchids.

ORCHID HABITATS | 11

Orchid-rich scrub in E Crete (25)

Orchid-rich evergreen oak forest in NW Sicily (25)

An abandoned field with *Orchis italica* in SC Cyprus (25)

An orchid-rich habitat in SC Crete

ORCHID HABITATS | 13

An orchid-rich habitat in C Crete (25)

Abandoned cultivated places are often rich in orchids, which are primary colonisers of such sites.

Some orchids flower very early in the year, with species such as *Himantoglossum robertianum*, *Orchis punctulata*, *Anacamptis collina* and *Ophrys bombyliflora* flowering in February or earlier in the warmest places.

All of the large islands of the central part of the Mediterranean are worth visiting to see orchids. The best islands are the largest, namely Sardinia, Corsica and Sicily. The smaller islands, such as the Balearics and Malta, have less diverse albeit interesting floras. In our experience, Sicily is one of the most rewarding places to visit because of its species richness and the low use of herbicides in agriculture and on the ancient sites.

Crete, Cyprus, Rhodes and the Aegean islands are amongst the richest orchid habitats and are often visited by orchidophiles. The floras of the Dodecanese and Cyprus have a distinctly Turkish influence. Thus, the range of *Himantoglossum comperianum* extends onto some of these islands from the Turkish mainland. *Orchis anatolica* and its close allies are found in Cyprus, Crete and Rhodes. Many *Ophrys* species, such as *O. reinholdii* also extend their ranges onto the Aegean islands. The ancient sites used to be amongst the best places to see a diversity of orchids. Thus, *Orchis simia*, *Limodorum abortivum* and a variety of *Ophrys* species could be found on the site at Olympia and on Mt Chronos adjacent to it. Delphi was also rich in orchids, notably *Ophrys*. Both sites, and many others have now been sprayed with herbicides that have almost eliminated the orchids and the rest of the flora. The same is true in Turkey and in mainland Italy.

A number of Turkish orchids are not found in Europe. These include *Cephalanthera caucasica* and *C. kurdica* and several *Ophrys* species, such as *O. cilicica* and *O. schulzei*. Salep-digging, the annual harvesting from the wild of the tubers of *Orchis*, *Ophrys* and *Himantoglossum* species among others, has led to many species being difficult to find in Turkey. An estimated 60,000,000 orchid tubers are harvested there annually. The prevalence of goats and sheep and the annual transmigration of herds into the mountains has also led to a decline in the orchids, which are preferentially cropped by these animals. In large parts of Anatolia, cemeteries now constitute havens for orchids because they are guarded from lifestock and are also generally left in peace by the salep collectors.

Salep on sale in the Istanbul spice market (25)

Montane grassland and woodland

Montane grasslands and meadows, especially those on calcareous substrates, can have a rich orchid flora that includes several specialities, such as *Chamorchis alpina*, *Nigritella rhellicani*, *N. miniata*, and *Traunsteinera globosa*, as well as a diversity of species that are also found at lower elevations.

As the spring develops, flowering orchids whose flowering period is already over at lower elevations can be sought in the hills and mountains of the Mediterranean, even into June and early July. Some orchids, especially species of *Cephalanthera*, *Epipactis* and *Limodorum*, prefer the shade of woodland. Open pine forests can also be good places for hunting orchids such as *Anacamptis morio*, *Orchis anatolica*, *O. provincialis* and *Dactylorhiza romana*.

Dactylorhiza sambucina growing in an alpine mea N Italy, Dolomites (21)

ORCHID HABITATS | 15

Orchids are often found amongst the ruins of ancient sites.
Greece, S Peloponnese (25)

Chamorchis alpina growing in the high Alps (21)

Grasslands and meadows

Dry meadows are increasingly rare throughout Europe as farmers plough and fertilise them. Those that remain are to be treasured and some are orchid-rich with species such as *Neotinea ustulata*, *Orchis militaris*, *Himantoglossum hircinum*, *Ophrys holosericea, O. sphegodes*, and *Anacamptis morio*.

Chalk and limestone grassland, meadows and open scrub in central and northern Europe provide a habitat similar to those in southern Europe, although the species composition is less rich. The typical spring orchid flora in central Europe includes *Anacamptis morio*, *A. pyramidalis*, *Dactylorhiza maculata* subsp. *fuchsii*, *Gymnadenia conopsea*, *Himantoglossum hircinum*,

Gymnadenia conopsea in a montane meadow, N Italy (21)

ORCHID HABITATS | 17

Orchids can be found in glades in open pine woodland, Majorca

A dry meadow near the Rhine, S Alsace, France, with numerous orchids (25)

Orchids growing in grassland studded with limestone boulders, Italy, Gargarno (21)

Neotinea ustulata, *Ophrys apifera*, *O. holosericea*, *O. sphegodes* and *Orchis militaris*. By early summer, *Coeloglossum viride*, *Gymnadenia densiflora* and *Herminium monorchis* are flowering, followed in late summer by *Spiranthes spiralis*, which can linger in flower until late September.

Cypripedium calceolus in deciduous woodland, Italy, Abruzzo (21)

Central and northern European woodlands

The deciduous woodlands of central and northern Europe, especially on chalk and limestone, have a characteristic spring orchid flora of *Cephalanthera damasonium*, *C. longifolia*, *C. rubra*, *Neottia ovata*, *Ophrys insectifera*, *Orchis pallens*, *O. purpurea* and *Platanthera chlorantha*. By July, *Epipactis helleborine*, *E. purpurata*, *E. atrorubens* and the self-pollinating *Epipactis* species begin to flower, some persisting in flower until September in the higher mountains. This is also the prime habitat of *Cypripedium calceolus*, the most charismatic European orchid, which grows in both deciduous and coniferous woodlands in dappled shade. In the Alps, it is often associated with spruce (*Picea abies*), but further north it is mostly found in ash (*Fraxinus*) and hazel (*Corylus avellana*) woods.

Dune slacks, fens and neutral to alkaline marshes

All of these wet, often calcareous, habitats can have a rich orchid flora. The early-flowering *Dactylorhiza incarnata*, in its many colour forms, is characteristic of these habitats. *Epipactis palustris* and *E. dunensis* flower a little later in the early summer. *Liparis loeselii* is also found in both dune slacks and fens, growing on wet moss clumps in the latter. *Malaxis monophyllos* favours neutral to slightly alkaline fens and bogs in coniferous woodland.

Bogs

Some *Dactylorhiza* species are found in acidic bogs. *Hammarbya paludosa* grows on *Sphagnum* moss clumps in oligotrophic bogs. *Neottia cordata* is often found in acidic conditions, in conifer forest, in *Sphagnum* moss, and amongst *Vaccinium* species.

The western sea-board

Although not rich in species, the western seaboard of north-west Europe can be one of the most rewarding places to hunt for orchids. It is the best place to look for *Hammarbya paludosa* and the only place in Europe to find *Spiranthes romanzoffiana*. Dactylorhizas grow in profusion in the meadows, which are called 'machair' in Scotland, and on the cliffs facing the Atlantic, and they include a few endemic subspecies or varieties of more widespread species.

Dune slacks in Northumberland, England, a rich habitat for orchids (22)

ORCHID HABITATS | 21

A bog in western Scotland: the habitat of *Hammarbya paludosa* (61)

Machair in western Scotland, a habitat rich in *Dactylorhiza* species (29)

Roadsides, gardens and waste places

Orchids will thrive in disturbed places, such as roadsides, until rank grasses and other vigorous plants outcompete them. Unusual habitats, such as the margins of gardens, gravel pits, power station slag heaps and infill and soil dumps, can be surprising habitats for orchids.

A lawn with a thriving colony of *Anacamptis morio* in Sussex, England (25)

CONSERVATION

People are nature's greatest enemy. We have had a profound impact on the landscape, especially in Europe where many orchid-rich areas have disappeared, often through change in land-use from pasture to arable use, urban development or deforestation. Not only are the orchids lost but their pollinators also suffer and become threatened.

Some of the most damaging changes are the result of over-use of fertilisers in meadows and the use of pesticides on arable land. This is particularly noticeable in many of the classical sites in the Mediterranean, which used to be amongst the best places to see orchids but have mostly been sprayed with herbicides to reveal the ruins better. Anthropogenic acid rain and climate change, road construction, and draining and peat-cutting in marshes and bogs also have profound detrimental effects on the habitats of orchids. Of course, Europe has many nature reserves and protected areas where orchids can be found but many species-rich areas have little or no protection. In these habitats, mowing and grazing should be restricted until after seed maturation.

European orchids are protected by local, national and international legislature. Unfortunately, these laws are often overlooked in areas lacking specific protection, particularly when land use is changed, for example, by enrichment by artificial fertiliser. Within the European Union, several orchids are listed in Annex II or IV of the *Habitats Directive*, meaning that any EU member state where one or more of these species occur is legally obligated to protect the species and their habitats.

Some farmers are aware of their responsibility to the environment and operate an ecologically sensitive approach to agriculture, for example, by dispensing with or reducing the use of pesticides. This supports habitats and helps those plants and pollinators that are threatened with extinction. Nevertheless, the number of good orchid habitats is decreasing rapidly. Too often, the environment is damaged for economic reasons to increase crop yields.

The life expectation of orchids is difficult to measure and differs between species. The average life-span of *Ophrys sphegodes* in England has been found to be eight years. By contrast, some plants of *Cypripedium calceolus* can survive for decades in the wild.

Orchis simia on a roadside verge in Umbria, Italy (21)

Reclaimed maize field seeded with locally sourced wildflower seed and subsequently colonised by several orchid species, Switzerland, Jura (25)

Orchids can survive for some years without flowering and will respond to the resumption of favourable conditions. This is quite usual and is dependent on factors such as climatic and environmental conditions at the time of flowering. Exceptionally, both *Limodorum abortivum* and *Epipogium aphyllum* have been found flowering below ground level in dry summers.

Awareness of the threats to orchids is rising and a number of initiatives have been taken to protect rare species and improve their numbers. Among the most exciting are attempts to achieve integrated conservation strategies for rarities. This can involve efforts to increase seed set and to propagate seedlings for eventual re-introduction, efforts to restore damaged habitats where orchid numbers have declined through changes in woody plant cover or invasion by alien species, and the creation of new orchid habitats. The last can be particularly effective because many orchids are primary colonisers of bare soil where competition from other plant species is removed.

One of the most promising developments in recent years has been the rise in eco-tourism, which benefits the local economy by encouraging groups of orchid enthusiasts to visit sites where they can photograph rare species. However, care must be taken not to damage the sites, for example, by compacting the soil around plants when taking photographs.

CLASSIFICATION

A great deal of new information has emerged in recent years about the age and relationships of the orchid family (Orchidaceae). The family is relatively ancient, having arisen some 100–120 million years ago (Givnish *et al.*, 2015). This estimate is based upon the assumption of a standard rate of mutation and calibrated using 17 fossils of flowering plants, including two of the very few known fossil orchids. The first confirmed orchid fossil is a pollinarium (pollen masses and attachment) found on the back of a ten-million-year-old bee fossilised in amber in the Dominican Republic (Ramírez *et al.*, 2007). Most other fossils ascribed to orchids lack critical features and their attribution is uncertain. By contrast, orchids are in active evolution and most of the 28,000 or more species that are currently accepted are probably of much more recent origin.

The orchid family can be subdivided into five distinct subfamilies: Apostasioideae, Cypripedioideae, Vanilloideae, Epidendroideae and Orchidoideae. Only species of three of them occur in Europe, namely, Cypripedioideae, Epidendroideae and Orchidoideae (see Table 1).

Table 1. European orchids belong in three of the five subfamilies currently recognised in the Orchidaceae

SUBFAMILY	TRIBE/SUBTRIBE	GENUS
Cypripedioideae		*Cypripedium*
Epidendroideae	Calypsoeae	*Calypso*
	Malaxideae	*Hammarbya, Malaxis, Liparis*
	Neottieae	*Cephalanthera, Epipactis, Limodorum, Neottia*
	Gastrodieae	*Epipogium*
Orchidoideae	Orchidinae	*Anacamptis, Chamorchis, Coeloglossum, Dactylorhiza, Gennaria, Gymnadenia, Gymnigritella, Habenaria, Herminium, Himantoglossum, Neotinea, Nigritella, Orchis, Ophrys, Platanthera, Ponerorchis, Pseudorchis, Serapias, Steveniella, Traunsteinera*
	Spiranthinae	*Spiranthes*
	Goodyerinae	*Goodyera*

Ophrys sphegodes subsp. *araneola*
France, Aveyron

TAXONOMY OF EUROPEAN ORCHIDS

The taxonomy and nomenclature of European orchids is confusing and a comprehensive scientific reassessment is long overdue. Any two authorities seldom agree as to how to classify and name them. In part, European orchids have suffered from their own popularity with the public, with a large fan-club of amateur enthusiasts and a number of specialist societies devoted to them across the continent. Many people, some with little formal scientific training, have added considerably to the literature on their taxonomy and nomenclature. First and foremost, confusion exists over what constitutes a genus, species, subspecies and variety. Even scientists are often unclear over the definition of these categories.

TAXONOMIC CONCEPTS

The genus

For nearly a century, generic concepts in European and Mediterranean orchids were stable and based predominantly on similarities in floral morphology. However, research undertaken in the 1990s and early in the present century, utilising DNA analysis, showed that a re-alignment was necessary. This research resulted in some species being transferred from one genus to another and to other genera being subsumed within others. This work is summarised in the second volume of *Genera Orchidacearum* (Pridgeon *et al.*, 2001). Thus, some species previously placed in the genus *Orchis* were shown to belong to *Anacamptis* and others to *Neotinea*, leaving a core of species in *Orchis*. Conversely, *Aceras* was placed back into *Orchis*, from which it had originally been separated because of its lack of a spur. Similarly, *Barlia* and *Comperia* were subsumed into *Himantoglossum*, *Neottianthe* into *Ponerorchis* and *Listera* into *Neottia*. Many of these changes resulted from the traditional classification being based predominantly on floral morphology. Unfortunately, these traits have proved less reliable in distinguishing taxa than was thought because pollinators exert a profound influence on flower structure. Thus, two unrelated species can evolve to have similar shaped and coloured flowers if pollinated by the same or closely related insects. By contrast, *Listera* was formerly distinguished from *Neottia* because of its distinctive lifestyle: species that were placed in *Listera* have green leaves and are photosynthetic, whereas those placed in *Neottia* are mycotrophic. However, their DNA shows that the holomycotrophic species evolved from those with well-functioning chlorophyll, and both are now included in *Neottia*.

The latest attempt to reconstruct the phylogeny of *Dactylorhiza* s.l. (as recognised in the first edition of this field guide), based on by far the most comprehensive genomic data to date, was published by Brandrud *et al.* (2020). Their results clearly indicate that *Coeloglossum viride* (*Dactylorhiza viridis* in the first edition) is sister to *Dactylorhiza* s.s. Thus, recognizing *Coeloglossum* as a separate monotypic genus is congruent with the phylogenetic relationships inferred, and we recognize it here based also upon its remarkable morphological distinctiveness from *Dactylorhiza*.

Similarly, Brandrud *et al.* (2019) reconstructed the overall phylogeny of *Gymnadenia* s.l., as recognized in the first edition of this field guide, based on the most comprehensive genomic data to date. Their results unequivocally indicate that *Nigritella* is sister to *Gymnadenia* s.s. Because of the remarkable morphological differences between them, we recognize *Nigritella* as a separate genus here. In line with this, we also recognize *Gymnigritella* as a separate genus, Brandrud *et al.* (2019) having reconfirmed that *G. runei* from Swedish Lapland (see first edition) has an allopolyploid origin from diploid *Gymnadenia conopsea* and tetraploid *Nigritella nigra*.

The species

Traditionally, taxa (a group of any rank) such as species, subspecies and variety have been defined morphologically. In other words, they have been distinguished from others at the same rank by morphological discontinuities. Thus, where morphological variation is continuous, only a single taxon has traditionally been accepted.

It needs to be stated that many species vary a lot. Take human beings as an example: not only is there sexual dimorphism but there are also differences in skin colour, eye colour and shape, hair type, body shape and many other features. Yet, all are interfertile and a single species is now accepted by all rational people. The same applies, in varying degrees, to plants of which orchids form a sizeable group (c. 8%). Variablility in orchids

Ophrys cretica flowers can vary even in the same spike

RIGHT
The variation in *Ophrys holosericea* subsp. *holosericea* in a single field in southern Alsace is truly remarkable

TAXONOMIC CONCEPTS

can be great even within a single population. An example is the variation found in a population of *Ophrys holosericea* shown on p. 29. Occasionally, even flowers in the same inflorescence can vary significantly from each other. Where morphology is not continuous, two or more taxa have been recognised. Unfortunately, the degrees by which species, subspecies and varieties can differ has seldom been defined. In general, species are usually considered to differ in a number of discrete characters, subspecies by fewer, and varieties by only one or two, e.g. flower colour or leaf shape.

The greatest contention amongst specialists working on European orchids is how to circumscribe the species. Two schools exist. In one, whose proponents are called 'splitters', species are narrowly defined, mostly on floral morphology, often backed up by suggestions that pollinator specificity is absolute (especially in *Ophrys*). The proponents of this approach look for morphological differences and place great emphasis upon them. In the other approach, adopted by those called 'lumpers', species are more broadly circumscribed and patterns of morphological variation that are continuous or more-or-less so are sought to define species.

Geographical factors are also usually considered. Are the morphologically different taxa sympatric (occurring in the same range) or allopatric (occurring discretely)?

Breeding behaviour has been widely used to distinguish allied species. Usually, such species are not inter-fertile or produce offspring with no or limited fertility and/or viability. Unfortunately, closely allied orchids are notoriously inter-fertile and often produce hybrids with high fertility. Fertility barriers often exist in orchids at the subtribal level, rather than at the level of species or even genus.

More recently, taxa have been established because of their presumed specific pollinator fauna, that is, the belief that each species has one or a few specific pollinators that preclude its crossing (hybridising) with a morphologically similar but not identical taxon. Unfortunately, the pollination of many taxa is incompletely understood, and information is frequently based on a few observations of insect visits, often without even confirmation of pollen transfer and subsequent fertilisation. We know that morphological characteristics of flowers and floral fragrances attract pollinators. Many detailed studies in the field show that a spectrum of insects are attracted to the flowers of most taxa, often with one insect species predominating at a particular locality. Furthermore, floral fragrance composition is known to be temperature dependent. Thus, we might expect different insects to be attracted under different temperature regimes and at different times of the day. When applied uncritically, assumed pollinator specificity is a poor criterion for species recognition.

Many European orchid taxa have been described several times and the plethora of names has led to immense confusion. This has often occurred because of a lack of knowledge of the literature, ignorance of the natural variation within a species or within a population, or a stamp-collecting philosophy that wants to recognise morphological difference no matter how trivial. Ignorance of the rules of nomenclature for plants (clearly set out in the *International Code of Nomenclature for Algae, Fungi and Plants*, which is updated every five years at the International Botanical Congresses) has also led to numerous invalid names, some of which have acquired a wide currency.

For the non-specialist, it is also difficult to understand why well-known scientific names are changed. This can be the result of changes in the rank of a taxon, the mis-interpretation of old names, the discovery of neglected ones or simply mis-identification.

A further factor has been the occasional occurrence of hybridisation between species, particularly in disturbed sites such as landslides and freshly made roadsides. This has resulted in orchid enthusiasts imagining hybridisation even when it has not occurred. Hybridisation does occur in European orchids and has been very well documented; for example, in the genera *Dactylorhiza*, *Ophrys* and *Orchis* and between genera such as *Gymnadenia* and *Dactylorhiza*. Polyploidy, usually in connection with hybridisation, has also been shown to lead to speciation in *Dactylorhiza*, usually in connection with hybridisation. This phenomenon represents a particular taxonomic challenge.

The recent analysis of the DNA of many European orchid taxa has reinforced our view that there has been an amazing degree of taxonomic inflation amongst them. For example, Delforge (2016) has recently listed 353 species of *Ophrys* as occurring in Europe and the Mediterranean region based on alleged differences in morphology and pollination syndrome. Yet, DNA analysis (Devey *et al.*, 2008) has shown that the number of discrete entities, from a genetic view-point, is less than a tenth of that number. How can these diametrically opposite views be reconciled?

The classification and nomenclature of *Epipactis* is also problematic. A number of species, for example, *E. palustris* and *E. veratrifolia*, are morphologically distinct. They are mainly out-crossing. However, other species, such as *E. helleborine* and its allies, are more difficult to define and name. *Epipactis helleborine* subsp. *helleborine* is also usually out-crossing, being pollinated mainly by wasps (*Vespula* spp.). A number of its allies have, however, lost their rostellum, allowing the pollen masses to ooze onto the stigma of the same flower, leading to self-pollination. Are these good species or merely insignificant forms and varieties? The recent DNA studies by Sramkó *et al.* (2019) have clarified some of the anomalies in *Epipactis* if not yet all of them. This has led to some reorganisation of the species and synonymy in this edition over that used in the first edition.

In the *Plants of the World Online* (Govaerts *et al.*, 2023), over 400 orchid taxa are recorded as occurring in Europe and over 500 are natural hybrids of European orchids.

This is, in our opinion, an over-statement of the true numbers. A re-assessment of morphology, and especially an understanding of natural variation, micro-morphology, chemistry and DNA analysis of the variation, has allowed us to rationalise this list here.

It is beyond the scope of this book to go into detail about the numerous scientific publications that have influenced our taxonomic views. However, in addition to the publications cited in the text, the 'Bibliography' includes a list of selected works from later years that have provided significant taxonomic guidance.

Subspecies, variety and form

These categories have been utilised to account for the natural variation seen within a species. The rank of subspecies is usually used where minor and consistent morphological differences between groups of populations are associated with geographical separation or with other obvious pre-pollination barriers, such as clearly different pollinator faunas. If no obvious pre-pollination barriers occur, then the category of variety is used. The rank of form is used for very minor differences, usually in flower colour, that occur in populations of typical plants.

Hybrids

Natural hybrids and hybrid swarms are a feature of the European and Mediterranean orchid floras. The reason is easy to fathom because closely allied species are often highly interfertile, being separated only by their habitat preferences, flowering time and pollinator specificity. When so much of the landscape in the region has been cultivated or otherwise disturbed by man, species that would usually be spatially, temporally or otherwise isolated come into close contact and hybridise more or less frequently. Many hybrids are occasional and probably not particularly long-lived. However, a number form significant populations over small to large areas. Their identity can be tricky to establish and often relies on the proximity of putative parents. In some cases, confusion can ensue. Further complication arises when primary hybrids backcross with either or both of the parental species, so that many second-generation and later hybrids resemble one or the other parent more closely. One result has been a proliferation of names, with hybrids often being described as new species, a feature that is particularly vexatious in *Ophrys*. Another complication arises when hybrids formed from the same parental species in different parts of their ranges (often involving different subspecies or varieties) are morphologically dissimilar, as has been seen in *Ophrys* and *Dactylorhiza*.

Hybridisation can eventually lead to the establishment of new species. In many cases, however, isolation is incomplete, especially where hybrid swarms are incompletely separated from either or both parents.

Hybrids can often be identified by having characteristics that are intermediate between those of their putative parental species, but a good knowledge of their morphology is necessary to confirm hybridisation. Particular attention should be paid to morphological analysis of the plant and to the habitat. With accurate studies, misdiagnosis can be reduced but not completely ruled out. Hybrids are frequently described as new species and provided with new names.

Primary hybrids are often taller, and larger in their parts, and have more brightly coloured flowers than in the parent species (the phenomenon known as heterosis). A good example is found in the intense and differently coloured flowers of hybrids between *Orchis anthropophora* and both *O. purpurea* and *O. militaris*; these flowers have colours that are not found in the parents. Striking flower colour is also found in hybrids between *Serapias* and *Orchis* and *Anacamptis*. Often the hybrids, for example, *Dactylorhiza maculata* subsp. *fuchsii* × *D. majalis* and *Orchis militaris* × *O. purpurea*, are more ecologically tolerant and survive longer.

The hybrids covered here fall into two categories:

Infrageneric hybrids arise when two species in the same genus cross. They can have two names: either a formula comprising the putative parental names separated by a multiplication sign, e.g. *Ophrys umbilicata* × *O. scolopax* and *Orchis militaris* × *O. purpurea*; or a distinctive new name with the genus name followed by a multiplication sign and a new epithet, such as *Dactylorhiza* ×*legrandiana* and *Ophrys* ×*vicina*. We provide the formula name for all the hybrids covered here and, where validated, the hybrid name and place of publication.

Intergeneric hybrids are hybrids between taxa in different genera and are usually rare. They can also be named using a formula, e.g. *Anacamptis laxiflora* × *Serapias vomeracea*, or with a hybrid generic name. The latter is formed by a multiplication sign preceding a compound generic name derived from a combination of the parental generic names, e.g. ×*Gymnanacamptis* (*Gymnadenia* × *Anacamptis*) or ×*Dactylodenia* (*Dactylorhiza* × *Gymnadenia*) and followed by a distinctive epithet.

The names for the hybrids are, in general, taken from the *Plants of the World Online* (Govaerts et al., 2023). We also use some names that are categorised as 'unplaced' there. For these, we have taken the name from *List of Hybrids of European Orchids* (Günther, 2023; www.guenther-blaich.de/hybnaminen.htm).

Epipactis palustris
Italy, Umbria (21)

HOW TO USE THIS FIELD GUIDE

This guide is predominantly a photographic one that aims to show, as far as possible, the variation found in each taxon (species, subspecies and variety) likely to be found in Europe across to Russia and the Caucasus, and in Mediterranean Europe and North Africa, including the Canary Islands, Madeira and the Azores. Identifications can be made by comparison of your plant with the images for each taxon. Dichotomous keys are provided to distinguish the genera and to identify the species within each genus in Appendix 1 at the end of the main text. As explained in the introduction, orchids, like most living organisms, vary in Nature. Often, no two plants growing in a population look quite alike. This should be borne in mind if your plant does not quite match anything shown here. It does not automatically follow that you have found a hybrid if your plant does not exactly match any of the images, although that possibility should be considered. We have included more images for particularly variable taxa than for those that vary less.

Dichotomous keys to the genera and species can be found in Appendix 1 on p. 419. Each couplet offers contrasting morphological states and lead either to a taxon or to the next couplet until identification is achieved.

The nomenclature largely follows that of the *Plants of the World Online* (Govaerts *et al.*, 2023). We list under 'Synonymy' all those additional names that are most commonly used in other current field guides and floras. The present range of each taxon is indicated under 'Distribution' and by an outline map (in both cases, only native occurrences are considered). The present ranges and distribution maps should be taken as a rough guide because no orchids are continuously distributed across their respective ranges, and because their ranges have been greatly affected by anthropogenic activity.

To assist identification, critical morphological features are given under 'Diagnostic features'. These should be taken as a rough guide, because occasional plants are larger or smaller than the typical ones. For example, *Epipogium aphyllum*, *Neottia nidus-avis* and *Limodorum abortivum* have been known to flower underground or when scarcely emerged in adverse conditions.

In addition to a list of publications cited in the text, the 'Bibliography' includes a list of other important systematic works – mainly recent studies with a relatively broad taxonomic or geographic scope – and a list of recommended regional orchid floras and excursion guides. We have given priority to recent books in English, French and German that contain detailed distribution maps and/or directions to interesting orchid sites.

Most of all, orchid hunting is a wonderful hobby. Enjoy it but please undertake it responsibly and do not damage the plants or habitat in pursuit of orchids.

GENERA AND SPECIES

1. CYPRIPEDIUM L., *Sp. Pl.* 2: 951 (1753)

Cypripedium is widely distributed across Europe, temperate Asia, and N America as far south as Honduras. Three species occur in Europe: *C. calceolus* and *C. macranthos* are found in calcareous, shady or semi-shady places, in deciduous and coniferous forests, and quite often on the margins of woodland; *C. guttatum* grows in more open habitats, often in light woodland, scrub or grassland.

The short creeping rhizomes have strong and fleshy roots. The stem bears 2–5 large, ovate, broadly lanceolate, plicate leaves. The floral bracts are similar to the foliage leaves. *Cypripedium calceolus* often has 2 or 3, and rarely 4, flowers whereas *C. guttatum* and *C. macranthos* are usually single-flowered. The lateral sepals are distinctively united and cup the back of the lip. The shoe-shaped lip is a trap for visiting insects. The column bears 2 lateral stamens and a terminal sterile stamen called a staminode. Insects that are lured to the staminode slip into the lip and have to crawl under the stamens and stigma to escape through the basal exits, thereby effecting pollination. Identification key on p. 422.

Cypripedium calceolus L., *Sp. Pl.* 2: 951 (1753).
Lady's slipper orchid

DISTRIBUTION. N and C Europe from N England and the Pyrenees to Russia, south to N Italy and the Balkans; also in northern temperate Asia across to Japan and a single occurrence in Algeria.

HABITAT. Shady woodlands, bushy hillsides, it needs partial shade; sea level to 2,000 m.

FLOWERING. May–July.

DISTINGUISHING FEATURES. 20–65 cm tall. Bears 1–4 flowers with maroon sepals and petals and a butter-yellow lip. Forms with pure yellow or yellow-green flowers are very occasionally found.

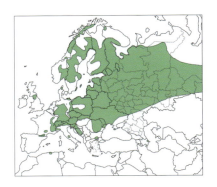

LEFT TO RIGHT FROM TOP
Germany, Black Forest 11.06.2014
Italy, Dolomites 26.06.2019 (25)
Germany, Black Forest 02.06.1983
Germany, Black Forest 06.06.1996
Germany, Black Forest 30.05.2005
Germany, Black Forest 01.06.1984
(note guide windows for the insect trapped in lip)
Finland, Kemi, Lapland 11.06.2014 (40)
Switzerland, Grisons (yellow form) 10.06.2014 (25)
Switzerland, Grisons (yellow form) 10.06.2014 (25)

Russia, Tomsk 22.06.2013 (46) Russia, Tomsk 22.06.2013 (46) Russia, Sakalin 01.07.2007 (46)

Cypripedium guttatum Sw., *Kongl. Vetensk. Acad. Nya Handl.*: 251 (1800). **Spotted lady's slipper**

DISTRIBUTION. European Russia and adjoining parts of Belarus and Ukraine; also across temperate Asia, south to the Himalayas and in Alaska.

HABITAT. Mixed deciduous and pine forest, grassland and open rocky areas; sea level to 4,100 m (2,700 m in Europe).

FLOWERING. May–June.

DISTINGUISHING FEATURES. 15–30 cm tall. Two almost opposite leaves on a distinct stem with a single small white flower heavily spotted with purple.

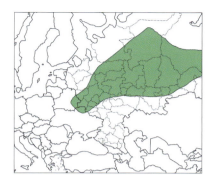

Cypripedium macranthos Sw., *Kongl. Vetensk. Acad. Nya Handl.*: 251 (1800). **Purple lady's slipper**

DISTRIBUTION. European Russia and an isolated outpost in Ukraine; also across northern temperate Asia to Korea, China, Taiwan and Japan.

HABITAT. Grassland, bushes, wet and shady places, river shores, preferring humus-rich soil; sea level to 2,400 m.

FLOWERING. June and July.

DISTINGUISHING FEATURES. Several-leaved plant with one large purple, pink or white concolorous flower.

NOTE. *Cypripedium macranthos* hybridises with *C. calceolus* where they are sympatric to form the hybrid *C. ×ventricosum*, which is usually purple-, pink- or white-flowered, differing from *C. macranthos* by its longer twisted petals.

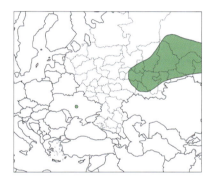

ALL IMAGES
Russia, Sakalin 27.06.2007 (46)

2. CEPHALANTHERA Rich., *Mém. Mus. Hist. Nat.* 4: 51 (1818).

Cephalanthera species have short creeping rhizomes and can be highly dependent on fungal nutrition, chlorophyll-less plants can occasionally be found. The lip, like that of *Epipactis*, is divided into a hypochile and a terminal epichile with longitudinal papillose or wavy ridges on its surface. The flowers of *Cephalanthera* are white or rose-purple. Pollination is autogamous or allogamous.

Cephalanthera species are found in open calcareous woodlands or rarely in dry grasslands. Identification key on p. 422.

Cephalanthera rubra (L.) Rich., *Mém. Mus. Hist. Nat.* 4: 60 (1818). **Red helleborine**

SYNONYMS. *Helleborine rubra* (L.) Schrank; *Limodorum rubrum* (L.) Kuntze

DISTRIBUTION. Throughout most of Europe and Turkey but often rare or local.

HABITAT. In deciduous, mixed and dry coniferous forest and in grassland, on calcareous soil, usually in shaded places; sea level to 2,400 m.

FLOWERING. May to July.

DISTINGUISHING FEATURES. 20–60 cm tall. Leaves grey-green, lanceolate, narrowly acute, 5–14 × 1–3 cm. Inflorescence axis and ovaries densely glandular-pubescent. Flowers rose-pink; sepals 17–25 mm long. Lip unspurred; epichile about twice as long as hypochile, with 7 yellow or green wavy ridges.

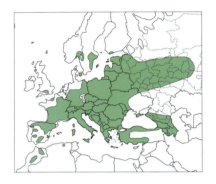

CLOCKWISE FROM TOP LEFT
Switzerland, Aargau 23.06.2001
Switzerland, Aargau 28.06.2005
France, Aveyron 30.06.2001
Switzerland, Aargau 23.06.2001
France, Auvergne 19.05.2011
Switzerland, Zurich Oberland 07.07.2007 (25)

Cephalanthera epipactoides Fisch. & C.A.Mey., *Ann. Sci. Nat., Bot. ser.* 4 (1): 30 (1854).
Spurred helleborine

SYNONYM. *Cephalanthera cucullata* subsp. *epipactoides* (Fisch. & C.A.Mey.) H.Sund.

DISTRIBUTION. E Greece, Turkey and Israel.

HABITAT. Particularly in oak and pine forest and scrub, on calcareous soils; sea level to 1,500 m.

FLOWERING. March to June.

DISTINGUISHING FEATURES. 20–70 cm tall. Flowers white with 7–9 cream to yellowish brown ridges on the narrow pointed epichile. Sepals 25–36 mm long; petals acute; lip with a 3–4 mm long spur.

BOTH IMAGES
Greece, Cos 09.04.1990 (1)

Crete 04.06.1986 (1)

Crete 04.06.1986 (1)

Crete 05.06.1987 (51)

Turkey, Gülnar
23.04.1990 (1)

Turkey, Gülnar
23.04.1990 (1)

Turkey, Antalya
3.05.2013 (40)

Turkey, Antalya
14.05.2013 (40)

Cephalanthera cucullata Boiss. & Heldr., *Diagn. Pl. Orient. Ser.* 1, 13: 12 (1854). **Cretan helleborine**
SYNONYM. *Epipactis cucullata* Labill.
DISTRIBUTION. Crete only.
HABITAT. In montane oak, maple, cypress and pine forest, on calcareous soil; 700–1,500 m.
FLOWERING. May and June.
DISTINGUISHING FEATURES. 15–30 cm tall. Similar to *C. epipactoides* but differs in having creamy white to pale pink flowers with obtuse petals and a spur that is only 1–2 mm long.

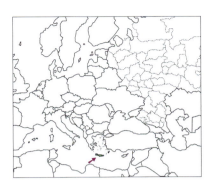

Cephalanthera kurdica Bornm. ex Kraenzl., *Bull. Herb. Boissier* 3 (3): 143 (1895). **Kurdish helleborine**
SYNONYMS. *Cephalanthera cucullata* subsp. *kurdica* (Bornm. ex Kraenzl.) H.Sund.; *C. cucullata* subsp. *floribunda* (Woronow) H.Sund.
DISTRIBUTION. Turkey, also in Iraq and Iran.
HABITAT. In montane holly and pine forest on calcareous soil; 500–2,100 m.
FLOWERING. April to June.
DISTINGUISHING FEATURES. 15–65 cm tall. Similar to *C. epipactoides* but differs in having pink to purple (very rarely white) flowers with 20–25 mm long sepals and a lip with an ovate, obtuse to rounded epichile.

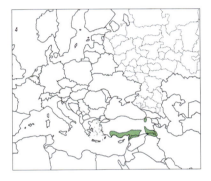

Cephalanthera longifolia (L.) Fritsch, *Oesterr. Bot. Z.* 38: 81 (1888). **Sword-leaved helleborine**
SYNONYMS. *Cephalanthera pallens* Rich.; *C. conferta* (B. &. H.Baumann) Kreutz
DISTRIBUTION. Most of Europe and temperate Asia, south to Israel, Pakistan and the Himalayas.
HABITAT. In deciduous forest and dry coniferous forest on calcareous soil, also in grassland, it prefers light shade; sea level to 1,200 m, rarely to 1,650 m.
FLOWERING. April to July.
DISTINGUISHING FEATURES. 20–70 cm tall. Leaves elongate and narrow, suberect to spreading, up to 18 cm long. Upper bracts vestigial. Flowers white with a yellow blotch on the epichile. Sepals 12–18 mm long; lip with 4–7 papillose ridges on the epichile, unspurred.

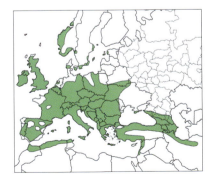

LEFT TO RIGHT
France, Aveyron 18.05.2011
France, Haute-Provence 29.05.2008
France, Aveyron 18.05.1991

Cephalanthera damasonium (Mill.) Druce, *Ann. Scot. Nat. Hist.* 1906: 225 (1906). **White helleborine**

SYNONYMS. *Cephalanthera ochroleuca* (Baumg.) Rchb.; *C. latifolia* Janch.

DISTRIBUTION. Most of Europe north to England and Gotland; also across temperate Asia.

HABITAT. Deciduous forest or rarely pine forest, dry and damp grassland usually in the shadow of bushes; sea level to 1,900 m.

FLOWERING. May–July.

DISTINGUISHING FEATURES. 20–60 cm tall. Uppermost bracts more than half as long as the ovaries. Flowers suberect, scarcely opening, white with an orange-yellow blotch on the epichile, self-pollinating; sepals 12–20 mm long; lip with 3–5 papillose ridges on the epichile, unspurred.

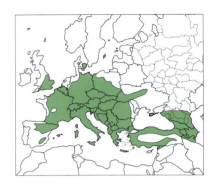

LEFT TO RIGHT FROM TOP
Italy, Abruzzo 01.06.2016 (21)
Turkey, Antalya 13.05.2013 (40)
France, Languedoc-Roussilon 22.05.1996
Germany, Black Forest 23.06.2005
France, Languedoc-Roussilon 18.05.1995
France, Haute-Provence 10.05.1998

Cephalanthera caucasica Kraenzl., *Repert. Spec. Nov. Regni Veg. Beih.* 65: 67 (1931). **Kotschy's helleborine**

SYNONYMS. *Cephalanthera pallens* Hohen., nom. illeg.; *C. kotschyana* Renz & Taubenheim; *C. damasonium* subsp. *kotschyana* (Renz & Taubenheim) H.Sund.

DISTRIBUTION. Turkey to the Caucasus and NW Iran.

HABITAT. Predominantly in oak forest and at forest edges, from sea level to 1,800 m.

FLOWERING. June and July.

DISTINGUISHING FEATURES. 20–70 cm tall. Similar to *C. damasonium*, but differs in having larger flowers (sepals over 20 mm long) that open to at least half-way.

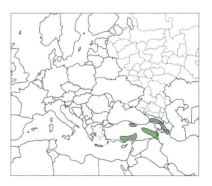

LEFT TO RIGHT
Turkey, Mersin 23.04.1990 (1)
Turkey, Mersin 23.04.1990 (1)

3. EPIPACTIS Zinn, *Cat. Pl. Hort. Gott.*: 85 (1757), nom. cons.

Epipactis is a large genus that is widespread in temperate Europe and Asia and extends into SE Asia and E Africa, with one species native in N America. Most species are found in woodland, forest margins and glades. In our region, the main exception is *Epipactis palustris*, which grows in swampy meadows, fens and moorland bogs, as well as in chalk grassland and dune slacks. Some species of *Epipactis* have a relationship with mycorrhizal fungi so strong that even occasional individuals that are devoid of chlorophyll may flower and survive for many years.

In recent years, numerous new species have been proposed. Most of them seem to represent taxonomically insignificant forms of already well-known species, but some appear morphologically intermediate and are difficult to place. The synonymies given here are only tentative, as further studies that are based on molecular and morphometric data are needed to improve our understanding of the complex variation patterns in *Epipactis*.

The flower structure is distinctive with the lip divided into a basal cup-shaped hypochile, which is usually filled with nectar, and a terminal flat or recurved epichile. In the column, the pollinia rest on the rostellum, the apex of which usually produces a drop of viscid matter surrounded by a membrane — a so-called 'diffuse viscidium'. The exact rostellum structure is closely related to the overall breeding system. Taxa with an effective and lasting viscidium depend on insects for pollination. Some of them are largely cross-pollinated (allogamous), whereas others mainly rely on pollinations among flowers in the same inflorescence (geitonogamy). Taxa that have an effective but evanescent viscidium are facultatively self-pollinated (autogamous); if insect pollination does not occur until the viscidium dries out, the pollinia will disintegrate and fall onto the stigma. Taxa in which the viscidium is ineffective or even absent are obligately autogamous. In some of the autogamous taxa, not only the viscidium but also the entire rostellum is more or less reduced. These taxa also frequently have partly closed flowers and a nectarless hypochile. Autogamy has evolved recurrently in *Epipactis*, implying that not all autogamous taxa are closely related.

The genus is related to *Cephalanthera*, which also has a bipartite lip, but species of the latter have white to rose-purple flowers that appear from April to July, whereas *Epipactis* species have duller flowers that usually begin to appear from July onwards.

We have largely followed Sramkó *et al.* (2019) and Bateman (2020) in their reassessment of the delimitation of taxa in European *Epipactis* using next generation DNA sequencing. Identification key on p. 423.

Epipactis veratrifolia Boiss. & Hohen., *Diagn. Pl. Orient. Ser.* 113: 11 (1913).
Eastern marsh helleborine

SYNONYM. *Arthrochilium veratrifolium* (Boiss. & Hohen.) Szlach.

DISTRIBUTION. Turkey and the E Mediterranean; also in the Arabian Peninsula across to the Himalayas and in NE Africa.

HABITAT. Slopes with seeping water, wet meadows and river banks, often in lime soil, sea level to 3,400 m.

FLOWERING. March to August.

DISTINGUISHING FEATURES. 20–150 cm tall. Easily recognised by the stout habit, the distinctly recurved petals, and the lip in which the shallow hypochile lacks both side lobes and nectar guides. The epichile is elastically attached, (ovate-)oblong and brownish with a white apex.

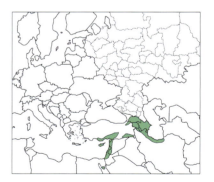

EPIPACTIS | 49

Cyprus 18.03.2002

Cyprus 14.03.2006

Cyprus 15.03.2011

Cyprus 15.03.2011

Epipactis palustris (L.) Crantz, *Stirp. Austr. Fasc.* (Ed. 2) 2 (6): 462 (1769). **Marsh helleborine**

SYNONYMS. *Helleborine palustris* (L.) Schrank; *Amesia palustris* (L.) A.Nelson & J.F.Macbr.

DISTRIBUTION. Widespread in Europe to N Anatolia and the Caucasus; also across temperate Siberia and Mongolia.

HABITAT. Damp and marshy meadows, dune slacks, marshes, by lakes and rivers; sea level to 2,200 m.

FLOWERING. June to August.

DISTINGUISHING FEATURES. Up to 65 cm, or more, tall, often forming extensive colonies because of the creeping and branching rhizome. Easily recognised by the mainly white petals and by the lip, which has linear nectar guides on the side lobes of the shallow hypochile. The epichile is elastically attached, suborbicular and white with a yellow spot.

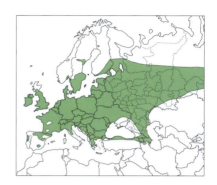

LEFT TO RIGHT FROM TOP
Switzerland, Obwalden 25.06.2002
Switzerland, Obwalden 25.06.2002
Switzerland, Obwalden 12.07.1999
Switzerland, Obwalden 22.06.2002
Switzerland, Obwalden 22.06.2002
Italy, Tuscany 04.07.2009 (21)

Epipactis atrorubens (Hoffm.) Besser, *Prim. Fl. Galiciae Austriac.* 2: 220 (1809). **Dark-red helleborine**

SYNONYMS. *Serapias latifolia* var. *atrorubens* Hoffm.; *Helleborine atrorubens* (Hoffm.) Druce

a. subsp. *atrorubens*

SYNONYMS. *Epipactis subclausa* Robatsch; *E. spiridonovii* Devillers-Tersch. & Devillers

DISTRIBUTION. From most of Europe across W Siberia to Tomsk, mostly in montane and hill forest except in the north.

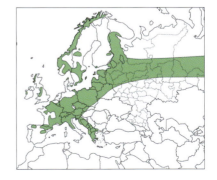

HABITAT. Open forest, on forest edges, scrub and dune slacks, on alkaline to neutral, nutrient-poor soil, and on well-drained sandy and stony ground; sea level to 2,400 m.

FLOWERING. June to August.

DISTINGUISHING FEATURES. 10–65 cm, or more, tall. Internode between inflorescence and uppermost leaf at least twice as long as any other internode; inflorescence axis pubescent. Leaves distichous; longest leaf more than 1.5 times as long as the internode above. Sepals more than 7 mm long (very rarely shorter); sepals and petals usually deep purple-red on ventral surface, rarely green or yellow. Epichile bearing prominent, strongly wrinkled-tuberculate calli. Viscidium effective and lasting. Ovary pubescent.

FROM BOTTOM OF P.50, LEFT TO RIGHT
subsp. *atrorubens*
Germany, Black Forest 13.07.1995
Switzerland, Aarau 18.07.2002
Germany, Black Forest 21.07.2002

OPPOSITE, LEFT TO RIGHT
Germany, Black Forest 21.07.2007
Switzerland, Grisons 19.07.2001
Switzerland, Grisons 21.07.1995
Switzerland, Grisons 28.06.2003
Switzerland, Grisons 13.07.1996
Switzerland, Grisons 13.07.2001

b. *Epipactis atrorubens* subsp. *parviflora*

A. & C.Niesch., *Philippia* 1 (2): 59 (1971).

SYNONYMS. *Epipactis parviflora* (A.Niesch. & C.Niesch.) E.Klein, nom. illeg.; *E. kleinii* M.B.Crespo, M.R.Lowe & Piera

DISTRIBUTION. Spain and SW France.

HABITAT. Open pine and oak forest; 700–1,500 m.

FLOWERING. May to July.

DISTINGUISHING FEATURES. 15–50 cm tall. Differs from the typical subspecies in having smaller flowers (sepals less than 6 mm long) and in the sepals and petals always being light green on their ventral surface.

BOTTOM ROW, LEFT TO RIGHT
subsp. *parviflora*
Spain, Galicia 13.06.1996 (1)
Spain, Burgos 06.2003 (13)
Spain, Burgos 06.2003 (13)

Germany, Baden-Württemberg
30.06.1984

Switzerland
18.09.1974

Germany, Black Forest
25.06.2003

Turkey, Boulou
25.06.1978 (1)

Turkey, Gümüshane
05.07.1978 (1)

Georgia 06.1979 (31)

Epipactis microphylla (Ehrh.) Sw., *Kongl. Vetensk. Acad. Nya Handl.*: 232 (1800).
Small-leaved helleborine
SYNONYMS. *Epipactis latifolia* var. *microphylla* (Ehrh.) DC.; *E. helleborine* var. *microphylla* (Ehrh.) Rchb.f.; *Amesia microphylla* (Ehrh.) A.Nelson & J.F.Macbr.
DISTRIBUTION. From C and S Europe across Asia Minor to Iran.
HABITAT. Shady bush and deciduous forest, rarely in coniferous forest, on calcareous soil; sea level to 1,900 m.
FLOWERING. May to August.
DISTINGUISHING FEATURES. 15–55 cm tall. Similar to *E. atrorubens* but differs in having remarkably small leaves (longest leaf less than 1.5 times as long as the internode above) and evanescent viscidia. Besides, the sepals and petals are never deep purple-red.

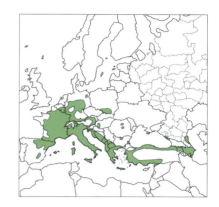

Epipactis condensata Boiss. ex D.P.Young, *Jahresber. Naturwiss. Vereins Wuppertal* 23: 106 (1970). **Eastern violet helleborine**
SYNONYMS. *Epipactis helleborine* subsp. *condensata* (Boiss. ex D.P.Young) H.Sund.; *E. kuenkeleana* (Akhalk., H.Baumann, R.Lorenz & Mosul.) P.Delforge
DISTRIBUTION. Transcaucasus, Crimea, Turkey, Cyprus, Israel, Syria and Lebanon.
HABITAT. Open montane pine and oak forest on strongly basic soil; 600–2,000 m.
FLOWERING. May to August.
DISTINGUISHING FEATURES. 20–80 cm tall. Vegetative parts light glaucous-green to yellow-green; inflorescence axis tomentose. Epichile bearing more or less verrucose calli. Viscidium effective but evanescent. Ovary pubescent to tomentose.
NOTE. *Epipactis krymmontana* Kreutz, Fateryga & Efimov, described from the Crimea, differs from typical *E. condensata* in being obligately autogamous. It might deserve taxonomic recognition at specific or infraspecific level.

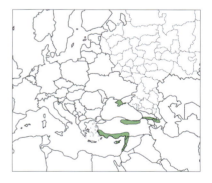

Epipactis pontica Taubenheim, *Orchidee (Hamburg)* 26: 68 (1975).

SYNONYM. *Epipactis persica* subsp. *pontica* (Taubenheim) H.Baumann & R.Lorenz

DISTRIBUTION. Scattered areas from N Italy and the Czech Republic to Georgia.

HABITAT. On neutral to slightly acidic soil in broadleaf forest, mainly beech forest; 500–1,500 m.

FLOWERING. June to August.

DISTINGUISHING FEATURES. 15–35 cm tall. Leaves dark to medium green, often somewhat clustered on mid-portion of stem. Inflorescence axis pubescent. Flowers only opening to about half-way (lateral sepals remaining parallel to lip). Petals and epichile never with pink or rose tones. Epichile smaller than hypochile, rounded to emarginate, at least as broad as long, bearing smooth to slightly furrowed calli. Viscidium ineffective. Ovary sparsely pubescent.

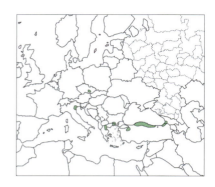

Epipactis helleborine (L.) Crantz, *Stirp. Austr. Fasc.* (Ed. 2), 2: 467 (1769). **Broad-leaved helleborine**

SYNONYM. *Serapias helleborine* L., nom. cons.

a. subsp. *helleborine*

SYNONYMS. *Epipactis latifolia* (L.) All.; *E. distans* Arv.-Touv.; *E. helleborine* var. *orbicularis* (K.Richt.) Verm.; *E. meridionalis* H.Baumann & R.Lorenz; *E. voethii* Robatsch; *E. aspromontana* Bartolo, Pulv. & Robatsch; *E. schubertiorum* Bartolo, Pulv. & Robatsch; *E. degenii* Szent. & Mónus; *E. molochina* P.Delforge; ?*E. autumnalis* Doro; *E. calabrica* U.Grabner, S.Hertel & Presser; *E. etrusca* Presser & S.Hertel; *E. sanguinea* S.Hertel & Presser; *E. collaris* S.Hertel; *E. lucana* Presser, S.Hertel & V.A.Romano

DISTRIBUTION. Widespread throughout Europe, Mediterranean NW Africa, and across Asia to the Himalayas. Naturalised widely in N America.

HABITAT. The most common *Epipactis*, in sunny and shaded places in deciduous forest and less often evergreen forest, at forest edges or in glades, on a variety of soils; sea level to 2,200 m.

FLOWERING. June to September.

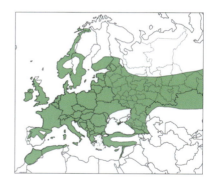

CLOCKWISE FROM LEFT
Switzerland, Aargau 13.07.2006
Switzerland, Solothurn 28.07.2001
Germany, Black Forest 23.06.2007
Denmark, Sjælland 26.07.2013 (45)

EPIPACTIS | 57

Turkey, Bursa-Kütahya
08.08.1996 (56)

Turkey, Bursa-Kütahya
08.08.1996 (56)

Turkey, Zonguldak
09.08.1996 (56)

Turkey, Zonguldak
09.08.1996 (56)

DISTINGUISHING FEATURES. 10–120 cm tall. Stem straight; inflorescence axis pubescent, thinner than ovaries. Leaves turning to all sides, dark to medium green, with an obtuse to rounded base and margins that usually are not wavy; longest leaf (8–)10–30 times as broad as the stem. Bracts in the mid-portion of the inflorescence pendent (to spreading), not distinctly broader than the floral lips; lowermost bract usually more than twice as long as the flower. Flowers opening more than half-way. Sepals and petals dull; petals and epichile sometimes with purple, pink or rose tones. Epichile at least as large as the hypochile, acute to retuse, at least as broad as long, bearing smooth to slightly furrowed calli. Viscidium present and usually effective; pollinia remaining intact, at least for some time after the flower has opened.

LEFT TO RIGHT, FROM TOP LEFT
subsp. *helleborine*
France, Alsace 29.06.2016
Switzerland, Aargau 18.07.2008
England, Surrey 28.06.1993 (21)
Denmark, Bornholm 01.08.2011 (45)

b. *Epipactis helleborine* subsp. *bithynica*

(Robatsch) Kreutz, *Kompend. Eur. Orchid.*: 61 (2004).
SYNONYM. *Epipactis bithynica* Robatsch.
DISTRIBUTION. Anatolia and Georgia.
HABITAT. 40–85 cm, or more, tall. Medium to deep shade in pine forest, beech woods, mixed forest and overgrown burial grounds, on neutral to alkaline soils; 700–1,800 m.
FLOWERING. Late May to August.
DISTINGUISHING FEATURES. Differs from the typical subspecies in having a more or less sinuous and fleshy stem (inflorescence axis thicker than ovaries) and in that the bracts of the mid-portion of the inflorescence are distinctly broader than the floral lips.

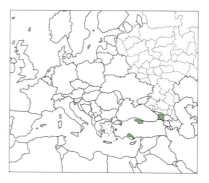

LEFT TO RIGHT, BOTTOM ROW
subsp. *bithynica*
Turkey, Antalya 22.05.2010 (45)
Turkey, Antalya 22.05.2010 (45)
Turkey, Antalya 22.05.2010 (45)

EPIPACTIS | 59

c. *Epipactis helleborine* subsp. *neerlandica*

(Verm.) Buttler, *Willdenowia* 16, 1: 115 (1986).

SYNONYMS. *Epipactis helleborine* var. *neerlandica* Verm.; *E. neerlandica* (Verm.) Devillers-Tersch. & Devillers

c.1. subsp. *neerlandica* var. *neerlandica*

SYNONYM. ?*Epipactis helleborine* var. *youngiana* (A.J.Richards & A.F.Porter) Kreutz

DISTRIBUTION. Coastal regions from Wales and northern France (Brittany, Normandy) across to Denmark and SW Norway.

HABITAT. In dune slacks in full sun or in mid-shade in willow scrub, less commonly in pine plantations and open woodland on sand dunes; near sea level.

FLOWERING. July to August.

DISTINGUISHING FEATURES. 7–55 cm or more tall. Distinguished from the typical subspecies in that the bracts of the mid-portion of the inflorescence are spreading to suberect at the time of flowering and in that the lowermost bract is less than twice as long as the flower.

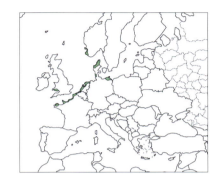

c.2. *Epipactis helleborine* subsp. *neerlandica* var. *renzii* (Robatsch) J.Claess., Kleynen & Wielinga, *Eurorchis* 10: 63 (1998)

SYNONYMS. *Epipactis renzii* Robatsch; *E. helleborine* subsp. *renzii* (Robatsch) Løjtnant; *E. neerlandica* var. *renzii* (Robatsch) P.Delforge

DISTRIBUTION. Northernmost Denmark.

HABITAT. In fully exposed coastal sand dunes, often among low willows; near sea level.

FLOWERING. July to August.

DISTINGUISHING FEATURES. An obligately autogamous taxon that differs from the typical variety in having flowers with a strongly reduced rostellum that never produces a viscidium. Besides, the flowers usually do not open widely.

OPPOSITE PAGE, TOP CLOCKWISE
subsp. *neerlandica*
Denmark, Tannisby 24.08.2001
Wales, Kenfig 07.1991 (22)
Wales, Kenfig 07.1991 (22)

OPPOSITE PAGE, BOTTOM LEFT TO RIGHT
subsp. *neerlandica* var. *renzii*
Denmark, Nordjylland 03.08.1998 (45)
Denmark, Nordjylland 31.07.1998 (45)
Denmark, Nordjylland 31.07.2007 (45)

d. *Epipactis helleborine* subsp. *tremolsii*

(Pau) E.Klein, *Orchidee (Hamburg)* 30: 49 (1979).

SYNONYMS. *Epipactis tremolsii* Pau; *E. atropurpurea* var. *tremolsii* (Pau) Schltr.; *E. helleborine* subsp. *latina* W.Rossi & E.Klein; *E. lusitanica* D.Tyteca; *E. turcica* Kreutz; ?*Epipactis cardina* Benito & C.E.Hermos.; *E. densifolia* W.Hahn, Passin & R.Wegener; *E. heraclea* P.Delforge & Kreutz; ?*E. helleborine* var. *castanearum* Gévaudan, Nicole & Anglade; *E. levantina* (Kreutz, Óvári & Shifman) P.Delforge

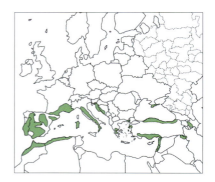

DISTRIBUTION. From the W and C Mediterranean region across Greece and Turkey to the Crimea, Azerbaijan and Israel.

HABITAT. Open pine, oak and holly forest, dry grassland, road edges and bush; sea level to 1,800 m.

FLOWERING. April to August.

DISTINGUISHING FEATURES. 20–90 cm, or more, tall. Differs from the typical subspecies in having (often somewhat leathery) leaves with more or less wavy margins and in the lowermost leaves, if not all leaves, having a (sub)cordate base. Besides, bracts of the mid-portion of the inflorescence are usually spreading at the time of flowering and often broader than the floral lips.

LEFT TO RIGHT FROM TOP
France, Hérault 22.05.2006
France, Hérault 22.05.2006
France, Hérault 22.05.2006
France, Hérault 27.05.2009
France, Hérault 22.05.2006
France, Hérault 24.05.2007

EPIPACTIS | 63

Epipactis purpurata Sm., *Engl. Fl.* 4: 41 (1828), nom. cons. **Violet helleborine**
SYNONYM. *Epipactis viridiflora* (Hoffm.) Krock.

a. subsp. *purpurata*

SYNONYMS. *Epipactis latifolia* subsp. *purpurata* (Sm.) K.Richt.; *Helleborine purpurata* (Sm.) Druce; *E. pseudopurpurata* Mered'a; *E. purpurata* var. *pollinensis* (B. & H.Baumann) P.Delforge

DISTRIBUTION. From England and Spain across C Europe to Lithuania, Italy, Greece, Moldova and possibly Ukraine.

HABITAT. Shady deciduous (beech, hornbeam) or rarely coniferous forest, mainly on clay; sea level to 1,600 m.

FLOWERING. July to September.

DISTINGUISHING FEATURES. 20–60 cm, or more, tall. Leaves distichous, grey-green, often flushed violet, lanceolate to narrowly elliptic or ovate; most, if not all, leaves longer than the average internodes; longest leaf 3–8(–10) times as broad as the stem. Inflorescence axis pubescent. Flowers opening fully. Sepals and petals shining like silk, light green to greenish white (petals very rarely with a faint rose tinge). Epichile obtuse to retuse, at least as broad as long, bearing slightly furrowed calli. Viscidium usually effective, but occasionally evanescent or even absent.

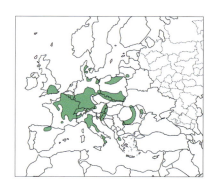

LEFT TO RIGHT
subsp. *purpurata*
Switzerland, Schaffhausen 03.08.2007
Switzerland, Schaffhausen 03.08.2007
Switzerland, Schaffhausen 01.08.2008
Switzerland, Schaffhausen 03.08.2007
subsp. *halacsyi*
Greece, Pelopponese 20.07.1991 (57)
Greece, Pelopponese 20.07.1991 (57)
Greece, Pelopponese 20.07.2001 (57)
Greece, Pelopponese 20.07.2001 (57)
Greece, Pelopponese 20.07.2001 (57)
Greece, Pelopponese 20.07.2001 (57)

b. *Epipactis purpurata* subsp. *halacsyi*

(Robatsch) Kreutz, *Ber. Arbeitskreis. Heimische Orchid.* 24 (1): 160 (2007).

SYNONYM. *Epipactis halacsyi* Robatsch

DISTRIBUTION. Endemic to the Peloponnese.

HABITAT. Shady broadleaf and coniferous forest on alkaline soils; 900–1,500 m.

FLOWERING. Late June to July.

DISTINGUISHING FEATURES. 20–40 cm tall. Distinguished from the typical subspecies by having broadly elliptic to (ob)ovate leaves that are dark green (sometimes flushed violet) and by the petals being rose to pink or, less frequently, light green suffused with pink.

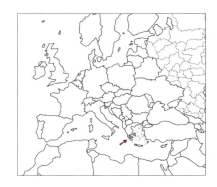

c. *Epipactis purpurata* subsp. *rechingeri* (Renz) Kreutz, *Eurorchis* 18: 94 (2006).

SYNONYM. *Epipactis rechingeri* Renz

DISTRIBUTION. Azerbaijan to N Iran.

HABITAT. Shady forest of beech or oak on shale or calcareous soil; 1,400–2,200 m.

FLOWERING. August and September.

DISTINGUISHING FEATURES. 20–70 cm tall. Distinguished from the typical subspecies by most, if not all, leaves being shorter than the average internodes and by the petals usually being suffused with pink.

LEFT TO RIGHT
subsp. *rechingeri*
Iran, Sang Deh, 21.08.1972 (51)
Iran, Sang Deh, 24.08.1973 (51)
Iran, Sang Deh, 25.08.1973 (51)
Iran, Sang Deh, 25.08.1973 (51)

Epipactis greuteri H.Baumann & Künkele, *Mitt. Arbeitskreis Heimische Orchid. Baden-Württemberg* 13: 344 (1981). **Greuter's helleborine**

SYNONYMS. ?*Epipactis nauosaensis* Robatsch; ?*E. olympica* Robatsch; *E. greuteri* var. *preinensis* (Seiser) P.Delforge; ?*E. leptochila* var. *komoricensis* (Mered'a) P.Delforge

DISTRIBUTION. C Europe, mainland Italy and across the Balkans to the Peloponnese.

HABITAT. Shaded and humid pine and deciduous forest on calcareous soils; up to at least 1,500 m.

FLOWERING. Late June to August.

DISTINGUISHING FEATURES. 20–80 cm tall. Leaves medium to dark green. Inflorescence axis pubescent; bracts pendent. Flowers pendent, opening to about half-way or rarely more. Epichile obtuse to acute, at least as broad as long, bearing smooth to slightly furrowed calli. Viscidium absent. Ovary distinctly curved; pedicel more than 5 mm long.

LEFT TO RIGHT
Italy, Campigna 29.05.2008 (57)
Greece, Pertouli 17.07.2001 (57)
Italy, Campigna 29.05.2008 (57)
Italy, Campigna 29.05.2008 (57)
Italy, Campigna 29.05.2008 (57)
Slovenia 15.08.1989 (57)

EPIPACTIS 67

Epipactis leptochila (Godfery) Godfery, *J. Bot.* 59: 146 (1921). **Narrow-lipped helleborine**
SYNONYMS. *Helleborine leptochila* (Godfery) Druce; *Epipactis helleborine* subsp. *leptochila* (Godfery) Soó; *E. muelleri* var. *leptochila* (Godfery) P.D.Sell

a. subsp. *leptochila*

SYNONYMS. *Epipactis leptochila* var. *dinarica* (S.Hertel & Riech.) P.Delforge; *E. leptochila* var. *futakii* (Mered'a & Potucek) P.Delforge; *E. maestrazgona* P.Delforge & Gévaudan; *E. leptochila* var. *peitzii* (H.Neumann & Wucherpf.) P.Delforge; ?*E. pinovica* S.Hertel, Tsiftsis & Z.Antonop.; *E. leptochila* var. *savelliana* (Bongiorno, De Vivo & Forsi) P.Delforge

DISTRIBUTION. From Britain and Denmark across C Europe to S France, Italy, the N Balkans and the Crimea.

HABITAT. Shady deciduous forest, or rarely pine forest, on calcareous soil; sea level to 1,550 m.

FLOWERING. Late June to August (about 2 weeks later than co-occurring subsp. *neglecta*).

DISTINGUISHING FEATURES. 15–70 cm, or more, tall. Leaves medium to dark green, turning to all sides. Inflorescence axis pubescent; bracts more or less spreading. Flowers spreading to nodding, opening more than half-way (rarely cleistogamous). Epichile porrect, acuminate, longer than broad, bearing smooth to slightly furrowed calli. Viscidium absent. Ovary straight, sparsely pubescent.

OPPOSITE, LEFT TO RIGHT
subsp. *leptochila*
Denmark, Møn 09.08.1998 (45)
Germany, Black Forest 28.07.1994 (cleistogamous form)
Germany, Black Forest 28.07.2004
Germany, Black Forest 21.07.1994
subsp. *neglecta*
Germany, Black Forest 25.07.1999
Germany, Black Forest 23.07.2003
Germany, Black Forest 21.07.2004
Germany, Black Forest 30.07.2003

b. *Epipactis leptochila* subsp. *neglecta* Kümpel, *Mitt. Arbeitskreis. Heimische Orchid. DDR* 11: 29 (1982).

SYNONYMS. *Epipactis neglecta* (Kümpel) Kümpel; *E. leptochila* var. *neglecta* (Kümpel) Gévaudan; *E. leptochila* var. *thesaurensis* (Agrezzi, Ovatoli & Bongiorni) P.Delforge & Gévaudan

DISTRIBUTION. C Europe south to Greece and Italy.

HABITAT. In medium to deep shade in beech and oak-hornbeam woodland, on calcareous soils; up to 1,500 m.

FLOWERING. June to early August (about 2 weeks earlier than co-occurring subsp. *leptochila*).

DISTINGUISHING FEATURES. 20–60 cm tall. Differs from the typical subspecies in having a recurved epichile and an ineffective viscidium.

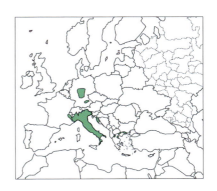

Epipactis dunensis (T. & T.A.Stephenson) Godfery, *J. Bot.* 64: 68 (1926). **Dune helleborine**

SYNONYMS. *Helleborine viridiflora* f. *dunensis* T. & T.A. Stephenson; *H. leptochila* var. *dunensis* (T. & T.A.Stephenson) Druce; *Epipactis bugacensis* Robatsch; *Epipactis rhodanensis* Gévaudan & Robatsch; *E. campeadorii* P.Delforge; *E. muelleri* var. *dunensis* (T. & T.A.Stephenson) P.D.Sell.; ?*E. provincialis* Aubenas & Robatsch; *E. sancta* (P.Delforge) P.Delforge; *E. dunensis* var. *tynensis* (Kreutz) P.Delforge; ?*E. maricae* (Croce, Bongiorni, De Vivo & Fori) Presser & S.Hertel

DISTRIBUTION. Britain, Spain across SE France to C and E Europe.

HABITAT. Sand dunes, willow scrub and pine reforestation in UK, deciduous forest, usually of beech or poplar, on low ground, often riparian in Continental Europe; sea level to 1,600 m.

FLOWERING. Late May to early September.

DISTINGUISHING FEATURES. 15–70 cm tall. Leaves distichous, yellow-green, suberect, straight or lightly recurved; longest leaf attached below the middle of the stem (inflorescence not counted in). Inflorescence axis pubescent. Flowers spreading, opening to about half-way; sepals and petals greenish-yellow to light green. Inner lateral walls of hypochile almost touching each other at their junction with the epichile, which is recurved, obtuse, at least as broad as long and provided with smooth to slightly furrowed calli. Viscidium absent. Ovary straight; pedicel less than 5 mm long.

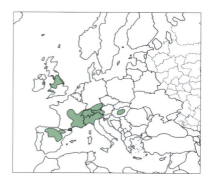

LEFT TO RIGHT FROM TOP
Wales, Fernby 22.07 1997 (3)
Wales, Anglesey 23.07.1991 (22)
Wales, Anglesey 21.07.1991 (1)
Switzerland, Solothum 04.06.1998 (8)
France, Lyon 28.06.1997 (57)
Switzerland, Geneva 27.06.1998 (8)

EPIPACTIS | 71

Epipactis muelleri Godfery, *J. Bot.* 59: 106 (1921).
Müller's helleborine

SYNONYMS. *Epipactis latifolia* var. *muelleri* (Godfery) Schltr.; *Helleborine muelleri* (Godfery) Bech.; *E. helleborine* subsp. *muelleri* (Godfery) O.Bolòs, Masalles & Vigo; *E. placentina* Bongiorno & P. Grünanger; *E. muelleri* subsp. *cerritae* M.P.Grasso; *E. placentina* var. *robatschiana* (Bartolo, D'Emerico, Pulv., Terrasi & Stuto) P.Delforge; *E. muelleri* var. *saltuaria* Kreutz.

DISTRIBUTION. C and S Europe, east to former Yugoslavia, and in the Crimea.

HABITAT. Dry warm forest margins and scrub, road- and track-sides; sea level to 1,600 m.

FLOWERING. June to August.

DISTINGUISHING FEATURES. 10–65 cm, or more, tall. Leaves strongly recurved, often with wavy margins. Inflorescence axis pubescent. Inner lateral walls of the hypochile well separated at their junction with the epichile, which is recurved, obtuse (to acute), at least as broad as long and provided with smooth to slightly furrowed calli. Viscidium absent. Ovary straight; pedicel less than 5 mm long.

Epipactis albensis Nováková & Rydlo, *Preslia* 50: 162 (1978).

SYNONYM. *Epipactis nordeniorum* Robatsch; *E. fibri* Scappat. & Robatsch; *E. tallosii* A.Molnár & Robatsch; *E. mecsekensis* A.Molnár & Robatsch; *E. moravica* Batoušek

DISTRIBUTION. C Europe and Italy.

HABITAT. Medium to deep shade in often riparian deciduous forest on acidic ground; up to 500 m.

FLOWERING. Mid-July to September.

DISTINGUISHING FEATURES. 10–30 cm, or more, tall. Leaves straight to lightly recurved; longest leaf attached above the middle of the stem (inflorescence not counted in). Inflorescence axis pubescent. Flowers spreading to nodding, opening to about half-way. Petals sometimes tinged with rose. Inner lateral walls of the hypochile almost touching each other at their junction with the epichile, which is porrect, acute, about as long as broad and provided with smooth to slightly furrowed calli. Viscidium absent or evanescent, never effective. Ovary straight; pedicel less than 5 mm long.

EPIPACTIS 73

France, Alsace
13.07.2008 (50)

Germany, Black Forest
20.07.1994

Germany, Black Forest
20.07.1994

France, Rhône-Alpes
01.10.2011 (50)

France, Rhône-Alpes
01.10.2011 (50)

France, Rhône-Alpes
11.10.2010 (50)

Epipactis phyllanthes G.E.Sm., *Gard. Chron.*: 660 (1852). **Green-flowered helleborine**

SYNONYMS. *Helleborine persica* Soó; *Epipactis troodi* H.Lindb.; *E. persica* (Soó) Hausskn. ex Nannf.; *E. phyllanthes* var. *degenera* D.P.Young; *E. phyllanthes* var. *pendula* D.P.Young; *E. phyllanthes* var. *vectensis* (T. & T.A.Stephenson) D.P.Young; *E. confusa* D.P.Young; *E. helleborine* subsp. *phyllanthes* (G.E.Sm.) H.Sund.; *E. persica* subsp. *gracilis* W.Rossi; *E. phyllanthes* var. *olarionensis* P.Delforge; *E. fageticola* (C.E.Hermos.) Devillers-Tersch. & Devillers; *E. exilis* P.Delforge; *E. taurica* Fateryga & Kreutz.

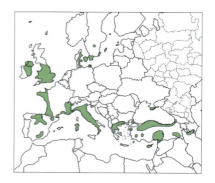

DISTRIBUTION. From Spain, SW France, the British Isles and S Sweden to mainland Italy and Sardinia, Greece, Cyprus, Turkey, the Crimea and the W Himalayas.

HABITAT. Mainly on low ground in shady deciduous forest, but also in coniferous forest, in scrub and on exposed sand dunes; sea level to 2,700 m.

FLOWERING. June to August.

DISTINGUISHING FEATURES. 10–70 cm tall. Leaves light green to dark green (sometimes flushed violet in the southern part of the range). Inflorescence axis subglabrous (hairs few and vestigial). Flowers nodding to pendent, opening to about half-way (rarely cleistogamous). Petals occasionally suffused with pink. Epichile recurved, obtuse to acute, about as long as broad, bearing smooth to slightly furrowed calli. Viscidium small and ineffective.

NOTE. The so-called *E. cretica*, described from Crete (but invalidly so, as no type specimen was cited), mainly differs from typical *E. phyllanthes* in having boat-shaped sepals and petals and an epichile with incurved margins. It should probably be recognised as a distinct species or as a subspecies of *E. phyllanthes*.

LEFT TO RIGHT FROM TOP
Denmark, Østiylland 06.07.1992 (45)
England, Northumberland 17.07.2002 (3)
Denmark, Lolland 30.07.2000 (45)
Crete 16.06.1985 (*Epipactis cretica*) (1)
Cyprus, Troodos 16.06.1990 (57)
Cyprus, Troodos 16.06.1990 (57)
Cyprus, Troodos 16.06.1990 (57)

4. NEOTTIA Guett., *Hist. Acad. Roy. Sci. Mém. Math. Phys.* (Paris, 4°) 1750: 374 (1754), nom. cons.

SYNONYM. *Listera* R.Br.

Three species of the *Neottia* are found in Europe. Two of them have a pair of opposite green leaves, between which rises a raceme of small green to brownish flowers. In the third species, the alternate leaves are reduced to sheaths, and the entire plant is pale brown, without visible chlorophyll. In all three species, the lip is cleft for at least one-third of its length, and the pedicel is distinctly set-off from the ovary. Identification key on p. 424.

Neottia cordata (L.) R.Rich., *De Orchid. Eur.*: 37 (1817). Lesser twayblade

SYNONYMS. *Ophrys cordata* L.; *Listera cordata* (L.) R.Br.; *Neottia nephrophylla* (Rydb.) Szlach.

DISTRIBUTION. Throughout N and C Europe; also across temperate Asia and N America.

HABITAT. In coniferous and birch forests and often in damp and shady places, especially in willow scrubs, and on moorland with *Vaccinium* and sorrel on acidic soils; sea level to 2,300 m.

FLOWERING. May–August.

DISTINGUISHING FEATURES. 10–20 cm tall. A tiny, often overlooked plant with green to dark purple flowers. Reminiscent of the markedly larger *N. ovata* but differs in having a subglabrous inflorescence axis and a lip with acuminate lobes and a pair of prominent lateral teeth at the base.

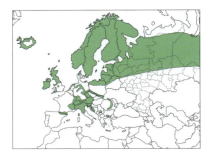

LEFT TO RIGHT FROM TOP
Switzerland, Obwalden 22.06.1995
Finland, Ou Lanka 12.06.2014 (40)
Switzerland, Obwalden 22.06.2005
Finland, Ou Lanka 05.06.2014 (40)
Switzerland, Obwalden 18.06.2008
Switzerland, Obwalden 18.06.2008

Neottia ovata (L.) Bluff & Fingerh., *Comp. Fl. German.* (Ed. 2) 2: 435 (1838). Twayblade

SYNONYMS. *Ophrys ovata* L.; *Listera ovata* (L.) R.Br.

DISTRIBUTION. Throughout most of Europe, especially in central and northern countries; also Asia, across to the Himalayas.

HABITAT. Woodland and glades and forest edges, also in grassland and meadows near woodland; sea level to 2,400 m.

FLOWERING. May to August.

DISTINGUISHING FEATURES. 20–60 cm tall. Distinguished by its 2 large opposite leaves borne well above the substrate, by its raceme with a densely glandular-pubescent axis and by bearing small, gnat-like, green to yellow-green flowers, the lip of which has rounded lobes.

LEFT TO RIGHT
Switzerland, Baselland 11.05.2016
Switzerland, Baselland 11.05.2016
France, Aveyron 22.05.1989
France, Aveyron 22.05.2007

Neottia nidus-avis (L.) Rich., *De Orchid. Eur.*: 37 (1817). **Bird's nest orchid**

SYNONYMS. *Ophrys nidus-avis* L.; *Serapias nidus-avis* (L.) Steud.

DISTRIBUTION. Throughout most of Europe; also in Asia across to Iran and Sakhalin.

HABITAT. In shade under beech and in deciduous and coniferous forests with rich humus on basic to neutral soils, preferring calcareous clay; from sea level up to 1,700 m.

FLOWERING. May to July.

DISTINGUISHING FEATURES. 15–40 cm tall. Fully mycotrophic and lacking leaves and visible chlorophyll; appearing above the ground only when it flowers. Its short rhizome has fleshy unbranched roots, looking like a bird nest. The flower lip is lowermost and has a nectar-filled basal cavity.

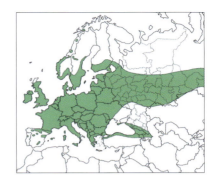

LEFT TO RIGHT
Switzerland, Obwalden 12.06.2017
France, Languedoc-Roussillon 18.05.2007
France, Languedoc-Roussillon 18.05.2007
Germany, Black Forest 12.06.1995

5. LIMODORUM Boehm., *Definitiones Generum Plantarum* (Ed. 3): 358 (1760).

Limodorum is a small genus confined to W & C Europe and the Mediterranean region east to Iran. It has a short stout rhizome, bearing numerous roots that run deep into the soil. The stem is steel blue to purple, rarely greenish white, the small sheathing leaves are of the same colour. Before flowering, the plant looks like an asparagus spear. The large flowers (lateral sepals more than 15 mm long) have an erect ovary and a long slender column, and they are usually self-pollinating. In a dry spring, *Limodorum* can flower and produce seeds underground. Prefers poor and calcareous soils, scrub, pine forest, oak woodland and roadsides. Identification key on p. 424.

Limodorum abortivum (L.) Sw., *Nova Acta Regiae Soc. Sci. Upsal.* 6: 80 (1799). **Violet limodore**

SYNONYMS. *Orchis abortiva* L.; *Jonorchis abortiva* (L.) Beck

DISTRIBUTION. Mediterranean and C Europe, Turkey and across to Iran; also in Mediterranean N Africa.

HABITAT. Pine forest, garigue, rarely in open oak and deciduous woodland, on base-rich and lime-rich soils; sea level to 1,900 m.

FLOWERING. April to July.

DISTINGUISHING FEATURES. 20–80 cm tall. Lip divided into a long-spurred hypochile and an elastically attached, elliptic to cordate epichile with incurved sides.

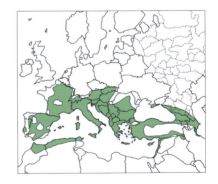

CLOCKWISE FROM TOP LEFT
Turkey, Antalya 19.05.2013 (40)
France, Haute-Provence 15.05.1985
Rhodes, Profitas Elias 07.06.2007 (25)
Rhodes 10.05.2007 (25)

Limodorum trabutianum Batt., *Bull. Soc. Bot. France* 33: 297 (1886). **Trabut's limodore**

SYNONYMS. *Limodorum abortivum* subsp. *trabutianum* (Batt.) Rouy, *L. abortivum* var. *trabutianum* (Batt.) Schltr.

DISTRIBUTION. Scattered in SW France and the Mediterranean region east to Greece.

HABITAT. Often in oak forest, beech- and coniferous forest (rarely in pine forest), on basic soils; sea level to 2,000 m.

FLOWERING. April to June.

DISTINGUISHING FEATURES. 20–55 cm tall. Lip narrowly lanceolate, flat, unspurred, not divided into hypochile and epichile.

6. EPIPOGIUM Gmel. ex Borgh., *Tent. Disp. Pl. German.*: 139 (1792).

A small genus of two species widespread in Europe, Asia and tropical Africa. They are leafless, chlorophyll-free geophytes with fleshy rhizomes and are fully reliant on fungal nutrition (mycotrophy) throughout their lifetime. The European species is able to survive and even flower underground and will not be noticed every year in most localities.

Epipogium aphyllum Sw., *Summa Veg. Scand.*: 32 (1814). **Ghost orchid**

SYNONYMS. *Epipogium generalis* E.H.L.Krause; *E. aphyllum* f. *albiflorum* Y.N.Lee & K.S.Lee

DISTRIBUTION. Throughout most of Europe but seldom in the Mediterranean zone and often very rare and local; also across temperate Asia south to the Himalayas. Britain's rarest orchid.

HABITAT. Shady places in humid beech and coniferous forest on nutrient-rich, basic soil; sea level to 1,900 m.

FLOWERING. July and August.

DISTINGUISHING FEATURES. 5–30 cm tall, often found in groups. The stem is bare, hollow and very fragile with brown scale leaves at the bottom. The 2–8 flowers are shortly stalked and are not resupinate, thus, the lip faces upwards. The sepals and petals are about 10–15 mm long. The 12–14 mm long lip is 3-lobed with 2 small, oblique side lobes and an ovate, concave epichile with 4–6 warty-papillose ridges. The spur is about 6–8 mm long.

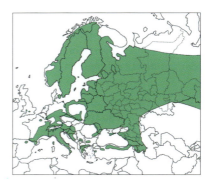

CLOCKWISE FROM TOP LEFT
Switzerland, Grisons 19.07.2002
Italy, Abruzzo 20.07.2013 (21)
Switzerland, Grisons 31.07.2001
Switzerland, Grisons 29.07.2003
Switzerland, Grisons 29.07.2003

Portugal, Sintra 24.04.1996 (1) Sardinia 17.05.1998 (1) Italy, Abruzzi 05.2002 (13)

7. HAMMARBYA Kuntze, *Revis. Gen. Pl.* 2: 665 (1891).

A single species with a circumboreal distribution. The tiny plants have a vertical rhizome, with a new pseudobulb produced every year above the dying one.

Hammarbya paludosa (L.) Kuntze, *Revis. Gen. Pl.* 2: 665 (1891). **Bog orchid**

SYNONYMS. *Ophrys paludosa* L.; *Malaxis paludosa* (L.) Sw.

DISTRIBUTION. N and C Europe as far south as S France; also across Asia and N America.

HABITAT. In fens and bogs in *Sphagnum* moss cushions on wet and poor, acidic, peat soil; sometimes on sandy soil in dune slacks. Owing to the draining of its habitat, this is an increasingly rare plant; sea level to 1,150 m.

FLOWERING. Late July and August.

DISTINGUISHING FEATURES. 5–20 cm tall. Of the usually 3 ovate, yellow-green leaves, the uppermost is largest, about 8–30 mm long and 5–10 mm wide. Propagules are produced from the leaf margins and these grow after they fall off. The inflorescence has up to 30–35 yellow-green non-resupinate flowers with recurved petals and a straight ovate acute lip. The bracts are at least half as long as the pedicelled ovaries.

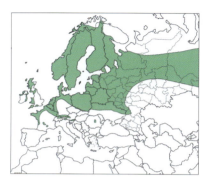

CLOCKWISE FROM LEFT
England, Hampshire 08.1987 (22)
Switzerland, Schwyz 15.07.1998
Switzerland, Schwyz 15.07.1988
(brood buds at the edge of the leaves)
Switzerland, Schwyz 03.08.1997

8. LIPARIS Rich., *De Orchid. Eur.*: 30 (1817).

Only a single species of this large cosmopolitan genus occurs in Europe. It is small plant with an underground pseudobulb.

Liparis loeselii (L.) Rich., *De Orchid. Eur.*: 38 (1817).
Fen orchid

SYNONYMS. *Ophrys loeselii* L.; *Orchis loeselii* (L.) MacMill.

DISTRIBUTION. From scattered areas in Europe to the central Russian uplands; also in N America.

HABITAT. Grows in moss in fens and swamps or on dune slacks by the sea, on basic substrates; sea level to 1,100 m. Threatened with extinction in many places because of changes in the land management and drainage of its habitat.

FLOWERING. June and July.

DISTINGUISHING FEATURES. 5–25 cm tall. Easily recognised by its usually 2 relatively large, light green, shiny, basal leaves and by its stem, which is triangular in cross-section and bears up to 15 greenish yellowish flowers. The abruptly recurved, rounded lip is also characteristic, as are the vestigial bracts. Plants are self-pollinating.

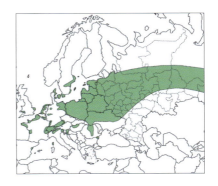

CLOCKWISE FROM LEFT
Switzerland, Obwalden 11.06.2002
Switzerland, Obwalden 17.06.2000
Switzerland, Obwalden 05.06.1995
Switzerland, Obwalden 12.06.2005
Switzerland, Obwalden 05.06.1995

Austria, Steiermark 06.07.2023 (45)　　Switzerland, Grisons 17.07.1998　　Switzerland, Grisons 15.07.2003

Finland, Lapland 01.06.2014 (40)　　Finland, Lapland 01.06.2014 (40)　　Sweden, Jokkmukk Pourtnak
　　　19.06.1995 (20)

9. MALAXIS Sol. ex Sw., *Prodr.*: 8, 119 (1788).

A single pseudobulbous species of this cosmopolitan genus occurs in Europe. The new year's pseudobulb grows from just above the last year's one.

Malaxis monophyllos (L.) Sw., *Kongl. Vetensk. Acad. Nya Handl.* 21: 234 (1800). **One-leafed bog orchid**

SYNONYM. *Ophrys monophyllos* L.

DISTRIBUTION. The Alps, northern and E Europe; also across Asia to Japan and Taiwan and in N America.

HABITAT. Grows in damp forest, in wet meadows or on roadsides, preferring semi-shade and calcareous soils; up to 1,900 m.

FLOWERING. June and July.

DISTINGUISHING FEATURES. 10–30 cm tall. It bears a single ovate to lanceolate leaf, rarely 2 or 3. The inflorescence can have up to 100 very small (2–3 mm across), insignificant, greenish and yellowish, non-resupinate flowers with slender sepals and petals and with a straight, cordate, long-acuminate lip. The bracts are at least half as long as the pedicelled ovaries.

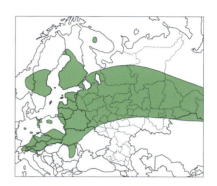

10. CALYPSO Salisb., *Parad. Lond.* t. 89 (1807).

A monotypic genus with a single species with a circumboreal distribution. Distinctive varieties are found in E Asia and in N America.

Calypso bulbosa (L.) Oakes, *Cat. Vermont Pl.* 28 (1842). **Calypso**

SYNONYMS. *Cypripedium bulbosum* L., *Calypso borealis* (Sw.) Salisb., *Orchidium arcticum* Sw.

DISTRIBUTION. Sweden, Finland and NW Russia; also in E Siberia, China, Japan and N America.

HABITAT. Boreal to subarctic moist coniferous forests; sea level to 800 m.

FLOWERING. May and June.

DISTINGUISHING FEATURES. 8–20 cm tall. Distinctive in having a single basal pleated stalked leaf and an erect inflorescence. The inflorescence bears a single pretty flower with pink sepals and petals and a white or pink-tinged lip, marked with yellow, and with a saccate base and 2 basal spurs.

11. CORALLORHIZA Gagnebin, *Acta Helv.* 2: 61 (1755).

Only a single species of *Corallorhiza* is found in Europe, and it is most common in the north and in the mountains of C Europe, the genus being more speciose in N America. Members of *Corallorhiza* are strongly mycotrophic, chlorophyll-deficient plants devoid of roots, but with coral-like rhizomes. They generally grow in shady, humus-rich forest, by dead tree trunks and in acidic peat, but also on basic soils.

Corallorhiza trifida Châtel., *Spec. Inaug. Corallorhiza*: 8 (1760). **Coralroot orchid**

SYNONYMS. *Ophrys corallorhiza* L.; *Neottia corallorhiza* (L.) Kuntze

DISTRIBUTION. Commonest in N Europe and in montane C Europe; also across Asia south to the Himalayas and in N America.

HABITAT. In beech, birch and moss-rich spruce forest, willow scrubs, tundra and bogs; sea level to 2,300 m.

FLOWERING. May to July.

DISTINGUISHING FEATURES. 8–30 cm tall. The slender stem bears a few leaf sheaths and a few greenish to brownish flowers with a white spurless lip usually spotted purple. The fruits are pendent.

LEFT TO RIGHT
Switzerland, Grisons 20.07.2009
Italy, South Tyrol 05.07.2003
Germany, Black Forest 05.07.2003
Finland, Ruka 05.06.2014 (40)
Italy, Trento 17.07.2010

12. GOODYERA R.Br. in W.T. Aiton, *Hortus Kew*. (Ed. 2) 5: 197 (1813).

A fairly large genus with a wide distribution, mainly in the Northern Hemisphere. Two species occur in Europe: *Goodyera repens* and *G. macrophylla*, the latter endemic to Madeira where it is threatened with extinction. These plants have creeping fleshy rhizomes, rooting at the nodes, and basal rosettes of leaves from which the often-glandular inflorescence arises. The flowers are small and do not open widely. The lip lacks a spur and has entire margins. Identification key on p. 424.

Goodyera repens (L.) R.Br. in W.T. Aiton, *Hortus Kew.* (Ed. 2) 5: 198 (1813). **Creeping lady's tresses**

SYNONYM. *Satyrium repens* L.

DISTRIBUTION. Widespread in Europe but avoiding the Mediterranean lowland; also in SW Asia to Azerbaijan and across temperate Asia to Korea and Japan, south to the Himalayas, and in N America.

HABITAT. In mossy pine and spruce forests and often with heaths in the Alps; sea level to 2,100 m.

FLOWERING. June to August.

DISTINGUISHING FEATURES. 5–25 cm, or rarely more, tall. An inconspicuous plant with dark green leaves that are up to 4.5 cm long and often reticulated with light green, with a 3–15 cm long inflorescence and small white glandular pubescent flowers.

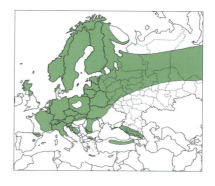

LEFT TO RIGHT
Italy, South Tyrol 19.07.2002
Switzerland, Grisons 01.08.1996
Switzerland, Grisons 03.08.2005
Switzerland, Grisons 03.08.2005
Switzerland, Grisons 01.08.1996

Goodyera macrophylla Lowe, *Trans. Cambridge Philos. Soc.* 4: 13 (1831). **Madeiran lady's tresses**

SYNONYM. *Peramium macrophyllum* (Lowe) Soó

DISTRIBUTION. Endemic to Madeira where it is very rare.

HABITAT. Wet and shady places, in steep laurel bush and forest; 800–1,400 m.

FLOWERING. September to November.

DISTINGUISHING FEATURES. 20–60 cm tall. Plant generally taller than *G. repens* with leaves that are more than 10 cm long when fully developed, a tapering, cylindrical inflorescence, up to 20 cm long, and more numerous flowers.

BOTH IMAGES
Madeira 20.09.2017 (59)

13. SPIRANTHES Rich., *De Orchid. Eur.*: 28 (1817).

A genus of c. 40 species, largely distributed in the Northern Hemisphere. Three species are considered to occur spontaneously in Europe. *Spiranthes aestivalis* and *S. spiralis* are relatively widespread but rare in C and W Europe. *Spiranthes romanzoffiana* is found in Ireland, W Scotland and the SW of England. It is also found in N America, which is the diversity centre of *Spiranthes*. The pink-flowered *Spiranthes australis* (R.Br.) Lindl. and the white flowered *S. cernua* (L.) Rich. × *S. odorata* (Nutt.) Lindl., *S. lucida* (H.H. Eaton) Ames and *S. odorata* have been found at a few sites. We consider them to be escapes from cultivation.

Spiranthes can be recognised by the glandular-hairy flowers that are arranged in a spiral or in vertical rows, and by the unspurred lip which has an erose front margin. Identification key on p. 424.

Spiranthes romanzoffiana Cham., *Linnaea* 3: 32 (1828). **Irish lady's tresses**

SYNONYM. *Spiranthes stricta* (Rydb.) A.Nelson

DISTRIBUTION. A very restricted distribution in Europe in Ireland, W Scotland, NW Wales and SW England; also in N America. The species has been recorded as an alien in The Netherlands.

HABITAT. Peaty pastures, flood plains and marshy meadows, by rivers and lake-shores; sea level to 200 m.

FLOWERING. July and August.

DISTINGUISHING FEATURES. 15–25 cm tall. It is a more robust plant than *S. aestivalis* and *S. spiralis* with much larger flowers. Besides, the flowers are arranged into 3 spirally twisted rows and have a lip that is obtuse to acute.

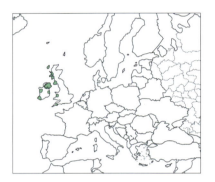

Spiranthes aestivalis (Poir.) Rich., *Mém. Mus. Hist. Nat.* 4: 58 (1818). **Summer lady's tresses**

SYNONYM. *Ophrys aestivalis* Poir.

DISTRIBUTION. C and SW Europe, also in NW Africa, increasingly rare as its habitats are destroyed. Extinct in England and the Channel Islands.

HABITAT. Calcareous fens, wet meadows and nutrient-poor marshes, in humic soils; sea level to 1,800 m.

FLOWERING. June to August.

DISTINGUISHING FEATURES. 10–35 cm tall. A small, slender and inconspicuous plant with a basal rosette of up to 6 erect, linear-lanceolate leaves. The inflorescence, arising from the centre of the rosette, is glandular at the top and bears 6–20 white flowers arranged into a single spirally twisted row. The flowers have a whitish to cream throat to the lip which is rounded at the front.

LEFT TO RIGHT
Switzerland, Bern 15.06.2006
Switzerland, Obwalden 03.07.1994
Switzerland, Bern 26.06.2007

SPIRANTHES | 91

Ireland, Lough Neagh 11.08.1984 (5)   Eire, Kerry 27.07.1984 (5)   England, Devon 03.08.1987 (26)

Switzerland, Aargau 12.09.1994   Switzerland, Aargau 12.09.1994 (50)   Switzerland, Aargau 12.09.1994

England, Hampshire 07.07.1987 (22)   Italy, South Tyrol 17.07.2010   Switzerland, Obwalden 28.06.200

Spiranthes spiralis (L.) Chevall., *Fl. Paris ii*: 330 (1827). **Autumn lady's tresses**

SYNONYMS. *Ophrys spiralis* L., *Spiranthes autumnalis* f. *parviflora* Soó

DISTRIBUTION. Widespread throughout Europe, except in Iceland, Scandinavia and large parts of E Europe; also in NW Africa and SW Asia.

HABITAT. In grassland and garigue, on lime-rich and calcareous soils, less commonly on neutral soils; sea level to 1,400 m.

FLOWERING. August to October.

DISTINGUISHING FEATURES. 15–35 cm tall. Similar to *S. aestivalis* but differs in flowering later, in having a rosette of much shorter, spreading leaves and in having a lip with a distinctly green throat.

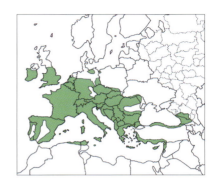

14. HERMINIUM L., *Opera Var.*: 251 (1758).

A single species is found in Europe, but more in Asia.

Herminium monorchis (L.) R.Br. in Aiton, *Hortus Kew.* (Ed. 2) 5: 191 (1813). **Musk orchid**

SYNONYMS. *Ophrys monorchis* L.; *Monorchis herminium* O.Schwarz

DISTRIBUTION. Widespread but very local in Europe and Asia across to China, Korea and Japan and south to the Himalayas. It has disappeared from many of its former localities.

HABITAT. Found in wet and dry meadows on calcareous soils and also in mountain meadows; sea level to 2,500 m.

FLOWERING. June to August.

DISTINGUISHING FEATURES. 8–30 cm tall. It forms colonies as new tubers arise from underground stolons. It has around 2–3, lanceolate to ovate leaves, bracts as long as the ovary and small and inconspicuous yellow flowers smelling of honey. The lip is hastate and the petals are longer than the sepals.

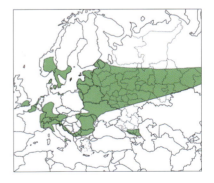

15. HABENARIA Willd., *Species Plantarum* (Ed. 4) 4 (1): 5, 44 (1805).

A large, almost cosmopolitan genus with a single species in our region. The genus is characterised by its terrestrial habit, spherical, ovoid or ellipsoidal tubers, green, either basal or cauline leaves and usually green, yellow or white flowers, usually with a 3-lobed lip and basal spur. Besides, it is the only European orchid genus that has the stigma placed on 2 freely extending branches, so-called stigmatophores.

Habenaria tridactylites Lindl., *Gen. Sp. Orchid. Pl.*: 318 (1835). **Canary Islands' habenaria**

SYNONYM. *Orchis tridactylites* (Lindl.) Webb & Berthel.

DISTRIBUTION. Canary Islands endemic.

HABITAT. Laurel forest, on wet and steep rocky lava slopes, also on derelict land and terraces; up to 1,400 m.

FLOWERING. November to February.

DISTINGUISHING FEATURES. 20–40 cm, or rarely more, tall. Readily distinguished by its yellow-green flowers, which have a lip that is almost completely divided into 3 narrowly linear segments.

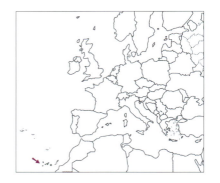

Canary Islands, La Palma
24.02.1996 (1)

Canary Islands, Tenerife
02.1995 (22)

Canary Islands, Tenerife
28.01.1995 (1)

16. GENNARIA Parl., *Fl. Ital.* 3, 2: 404 (1860).

A genus comprising only two species: one in S Asia and one in Macaronesia and the W Mediterranean.

Gennaria diphylla (Link) Parl., *Fl. Ital.* 3, 2: 405 (1860). **Gennaria**

SYNONYMS. *Coeloglossum diphyllum* (Link) Fiori & Paol., *Orchis diphylla* (Link) Samp.

DISTRIBUTION. Canary Islands, Madeira and the W Mediterranean (including NW Africa) east as far as Elba, Sardinia and Corsica.

HABITAT. Wet laurel forest of Madeira, on the Canaries in laurel and pine forest, often on margins in light shade, in the W Mediterranean near the coast in pine woodland or scrub; sea level to 1,300 m.

FLOWERING. March and April.

DISTINGUISHING FEATURES. 10–30 cm tall. The two alternate heart-shaped leaves and many tiny yellow flowers in a secund inflorescence are quite distinctive. Can form large colonies.

Canary Islands, La Palma
05.03.2016 (25)

Sardinia 03.2001 (23)

Elba 25.03.2009 (21)

17. PONERORCHIS Rchb.f., *Linnaea* 25: 227 (1852).

SYNONYM. *Neottianthe* (Rchb.f.) Schltr.

A predominantly Asiatic genus with many species in China and the Himalayas. Only a single species in Europe. Tubers ovoid to spherical.

Ponerorchis cucullata (L.) X.H.Jin, Schuit. & W.T.Jin, *Molec. Phylogen. Evol.* 77: 51 (2014). **Neottianthe**

SYNONYMS. *Orchis cucullata* L.; *Neottianthe cucullata* (L.) Schltr.

DISTRIBUTION. Poland, Latvia, Lithuania east to China, Korea and Japan and south to the Himalayas.

HABITAT. Sandy and moss-rich damp coniferous forest; sea level to 4,000 m (to 500 m in Europe).

FLOWERING. July–August.

DISTINGUISHING FEATURES. 10–30 cm tall. A slender plant with 2 basal leaves and a lax, secund, few-flowered inflorescence. Flowers pink, delicate with a 3-lobed lip with narrow lobes and a short, 5–6.5 mm long cylindrical spur.

BOTH IMAGES
Poland, Augustow 14.08.1999 (1)

---

18. PSEUDORCHIS Ség., *Pl. Veron.* 3: 254 (1754).

SYNONYM. *Leucorchis* E.Mey.

A genus comprising only a single species distributed in N Eurasia and N America. Related to *Gymnadenia* and similar in having two deeply divided and flattened tubers. The 4–8 broadly lanceolate leaves are unspotted and suberect. The inflorescence is cylindrical and densely flowered. Flowers are small and vanilla-scented with connivent sepals and petals and a short, 3-lobed lip with a short spur that is constricted at the base.

Pseudorchis albida (L.) Á.Löve & D.Löve. *Taxon* 18: 312 (1969). **Small white orchid**

SYNONYMS. *Satyrium albidum* L.; *Leucorchis albida* (L.) E.Mey.

a. subsp. *albida*

DISTRIBUTION. Most of Europe but absent from the Mediterranean lowlands, Iceland and the high mountains; also in north temperate Asia across to Kamchatka.

HABITAT. In hay meadows, grassland, dwarf hazel scrub and low fenland on acidic and calcareous soils; occurring from the lowland to the subalpine zone.

FLOWERING. May to August.

DISTINGUISHING FEATURES. 10–30 cm tall. Flowers greenish white, with 2–3 mm long sepals and petals and a 3–4 mm long lip, the mid-lobe of which is longer than the side lobes.

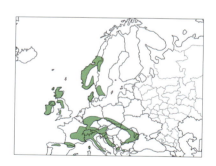

CLOCKWISE FROM TOP LEFT
Italy, South Tyrol 14.07.2009
Switzerland, Grisons 14.07.2009
Norway, Sør-Trøndelag 08.07.2004 (45)
Switzerland, Grisons 14.07.2009
Norway, Sør-Trøndelag 10.07.2004 (45)

Iceland, Vestur-Ísland 17.07.2006 (45)   Iceland, Norur-Ísland 19.07.2006 (45)   Iceland, Norur-Ísland 20.07.2006 (45)

b. *Pseudorchis albida* subsp. *straminea*

(Fernald) Á.Löve & D.Löve, *Taxon* 18 (3): 312 (1969).

SYNONYMS. *Habenaria albida* var. *straminea* (Fernald) F.J.A.Morris & E.A.Eames; *Pseudorchis straminea* (Fernald) Soják; *Gymnadenia albida* subsp. *straminea* (Fernald) Løjtnant; *G. straminea* (Fernald) P.Delforge; *Pseudorchis albida* f. *straminea* (Fernald) O.Gruss & M.Wolff

DISTRIBUTION. From Scandinavia to NW Russia, in the C European mountains, Iceland and the Faroe Islands; also in Greenland and E Canada.

HABITAT. Mainly in *Dryas* heaths but also in bristle grassland on calcareous soil; occurs in the subalpine and alpine zones, to 2,700 m.

FLOWERING. May to August.

DISTINGUISHING FEATURES. 10–30 cm tall. Differs from the typical subspecies in having straw yellow flowers with 3–4 mm long sepals and petals and a 4–5 mm long lip, the mid-lobe of which is about as long as the side lobes.

LEFT TO RIGHT
Greenland 18.07.1995 (16)
Sweden, Lycksele Lappmark 18.07.2005 (45)
Norway, Sør-Trøndelag 12.07.2004 (45)

---

19. PLATANTHERA Rich., *De Orchid. Eur.*: 20, 26, 35 (1817).

A large genus widely distributed in Europe, N Africa, Asia and N America. Ten species are found in our area but only two, *Platanthera bifolia* and *P. chlorantha*, are widespread.

Plants have turnip-shaped tubers, 1–3 basal, ovate to broadly lanceolate, unspotted leaves, a cylindrical and many-flowered inflorescence, and white, yellowish white or greenish flowers. The lip is entire, tongue-shaped and usually deflexed while the spur is slender and elongate. In the common species, the flowers smell of vanilla in the evening and are pollinated by night-flying moths. Identification key on p. 425.

Platanthera hyperborea (L.) Lindl., *Gen. Sp. Orchid. Pl.*: 287 (1835). **Iceland butterfly orchid**

SYNONYMS. *Orchis hyperborea* L.; *Limnorchis hyperborea* (L.) Rydb.

DISTRIBUTION. Iceland and Greenland. Whether it also occurs in Canada and the USA remains to be clarified.

HABITAT. Grassy hills, dwarf heath scrub, and grazed bush with good water supply; sea level to 100 m.

FLOWERING. July and August.

DISTINGUISHING FEATURES. 5–20 cm, or more, tall. Leaves 2–5, alternate on the basal part of the stem, broadest below the middle. Inflorescence dense, with small self-pollinating, dull cream to greenish white flowers having a narrow lip with a 3–6 mm long spur.

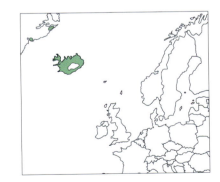

Platanthera oligantha Turcz., *Bull. Soc. Imp. Naturalistes Moscou* 27 (2): 86 (1854).
Northern butterfly orchid

SYNONYM. *Lysiella obtusata* subsp. *oligantha* (Turcz.) Tolm.

DISTRIBUTION. Found in northernmost Scandinavia and N Russia.

HABITAT. Birch wood regions, in subalpine and subarctic dwarf shrub heather; sea level to 1,200 m.

FLOWERING. June and July.

DISTINGUISHING FEATURES. 6–20 cm tall. A few-flowered species with a single basal leaf. The small flowers have 2–3 mm long sepals and petals and a slightly longer lip with a 2–3 mm long spur.

OPPOSITE, LEFT TO RIGHT
Sweden, Jayreoaivit 04.07.1976 (5)
Sweden, Torne Lappmark 11.07.2005 (45)
Sweden, Torne Lappmark 12.07.2005 (45)

Platanthera bifolia (L.) Rich., *Mém. Mus. Hist. Nat.* 4: 57 (1818). **Lesser butterfly orchid**

SYNONYMS. *Orchis bifolia* L., *Gymnadenia bifolia* (L.) G. Mey.

a. var. *bifolia*

SYNONYM. *Platanthera bifolia* subsp. *latiflora* (Drejer) Løjtnant

DISTRIBUTION. Insufficiently known but encompasses at least Atlantic NW Europe, the Alps, mainland Portugal and C Italy.

HABITAT. In sun or light shade in grassland, dune slacks, heaths, meadows and fens, on acid to base-rich soil; sea level to 2,500 m.

FLOWERING. June to July.

DISTINGUISHING FEATURES. 10–30(–45) cm tall. Leaves basal, usually 2 in number. Most internodes in the mid-portion of the spike 2–5(–7) mm long. Flowers white to cream with a 12–21(–25) mm long spur to the lip. Anther locules parallel and placed close together.

OPPOSITE, LEFT TO RIGHT
Platanthera bifolia
var. *bifolia*
Denmark, Nordjylland 15.06.2020 (45)
Denmark, Sjælland 27.06.2020 (45)
Denmark, Sjælland 27.06.2021 (45)
Denmark, Sjælland 27.06.2021 (45)
Denmark, Nordjylland 15.06.2020 (45)
Norway, Møre og Romsdal 15.07.2004 (45)

b. *Platanthera bifolia* var. *latissima* (Tinant) Thielens, *Bull. Soc. R. Bot. Belg.* 12: 99 (1873)

SYNONYMS. *Orchis bifolia* var. *latissima* Tinant; *Platanthera kuenkelei* H.Baumann; *P. fornicata* (Bab.) Buttler; *P. kuenkelei* var. *sardoa* R.Lorenz; *P. atropatanica* (H.Baumann, B.Baumann, R.Lorenz & Ruedi Peter) P.Delforge

DISTRIBUTION. From Europe (excl. Iceland, the Faroes, the British Isles and Macaronesia) to N Africa (Algeria, Tunisia), SW Asia and Mongolia.

HABITAT. In partial shade in open woodland and scrub, less often in full sun (e.g., in subalpine meadows), on base-rich soil; sea level to 2,200 m.

FLOWERING. May to June.

DISTINGUISHING FEATURES. 25–90 cm tall. Differing from the typical variety in the spur being usually > 25 mm long and in most internodes in the mid-portion of the spike being (4–)6–25 mm long.

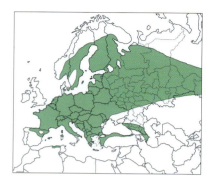

CLOCKWISE FROM TOP LEFT
Denmark, Østjylland 03.06.2019 (45)
Denmark, Østjylland 03.06.2019 (45)
Spain, Navarra 12.05.2014 (45)

Platanthera muelleri A.Baum & H.Baum, *J. Eur. Orch.* 49: 137 (2017). **Müller's butterfly orchid**

SYNONYM. *Platanthera bifolia* subsp. *osca* R.Lorenz, Romolini, V.A.Romano & Soca

DISTRIBUTION. Insufficiently known but encompasses at least parts of the Netherlands, Belgium, France, Germany, Switzerland, Austria and Italy.

HABITAT. In sun or light shade in grassland, meadows, scrub and open forest, on base-rich soil; lowland.

FLOWERING. May to July.

DISTINGUISHING FEATURES. 20–70 cm tall. Similar to *P. bifolia* but differs in having distinctly separated anther locules. This species is difficult to distinguish from the hybrid *P. bifolia* × *P. chlorantha* but it generally has a higher spur/lip length ratio (usually > 2.5 vs. < 2.5 in the hybrid). Unlike the hybrid, *P. muelleri* forms populations.

CLOCKWISE FROM THE LEFT
Belgium, Namur 18.06.2016 (56)
Germany, Rheinland-Westphalen 30.05.2020 (56)
Belgium, Namur 12.06.2015 (56)
Germany, Rheinland-Westphalen 30.05.2020 (56)

Platanthera chlorantha (Custer) Rchb., *Handb. Gewächsk.* (Ed. 2) 2: 1565 (1829).
Greater butterfly orchid

SYNONYMS. *Orchis chlorantha* Custer; *Platanthera bifolia* subsp. *chlorantha* (Custer) Rouy.

DISTRIBUTION. Widespread in Europe; also in SW Asia.

HABITAT. Open coniferous and deciduous forest, woodland margins, grassland and alpine meadows on a variety of soils; sea level to 2,500 m.

FLOWERING. May to August.

DISTINGUISHING FEATURES. 20–80 cm tall. Leaves basal, usually 2 in number. At least the lateral sepals white to cream. Lip straight to slightly recurved; spur 13–41 mm long, with a bilaterally flattened, slightly club-shaped apex. Anther loculi not parallel but forming an upturned V-shape, widest at the base by the viscidia. Differs from *P. holmboei* in having larger creamy to white flowers with 9–12 mm long, 5–6 mm wide lateral sepals, a 9–18 mm long lip, and an 18–41 mm long spur.

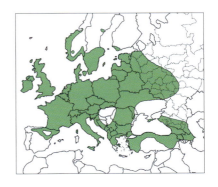

CLOCKWISE FROM INSET
France, Cantal 27.05.2017
France, Cantal 27.05.2017
Germany, Black Forest 12.06.2006
France, Massif Central 29.05.2002

Platanthera holmboei Lindberg f., *Årsbok Soc. Sci. Fenn.* 20B, 7: 5 (1942). **Holmboe's butterfly orchid**

SYNONYM. *Platanthera chlorantha* subsp. *holmboei* (Lindberg f.) J.J.Wood; *P. lesbiaca* Devillers-Tersch., Devillers, Dedroog, Baeten & Flausch.

DISTRIBUTION. Lesbos, Cyprus and Turkey to Israel.

HABITAT. In pine woods and scrub; up to 2,000 m.

FLOWERING. April to July.

DISTINGUISHING FEATURES. Similar to *P. chlorantha* but differing in having green to yellow-green lateral sepals and a spur that is not bilaterally flattened in its apical part.

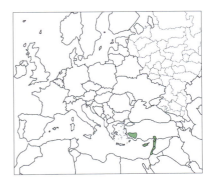

BOTH IMAGES
Cyprus 25.04.2004 (1)

Platanthera algeriensis Batt. & Trab., *Bull. Soc. Bot. France* 39: 75 (1892).
Green butterfly orchid

SYNONYM. *Platanthera chlorantha* subsp. *algeriensis* (Batt. & Trab.) Emb.

DISTRIBUTION. Corsica, Sardinia, S Spain and Mediterranean NW Africa.

HABITAT. Slope bogs, creek banks on high montane and subalpine places; 700–2,000 m. Previously also in coastal marshes.

FLOWERING. May and June.

DISTINGUISHING FEATURES. 30–70 cm tall. Similar to *P. chlorantha* but differing in having pale green to yellow-green lateral sepals. Furthermore, it has a strongly recurved lip and anther loculi that are slightly sigmoid – two features that also separate it from the even more similar *P. holmboei*.

BOTH IMAGES
Morocco, Ifrane 30.05.1996 (1)

Platanthera azorica Schltr., *Repert. Spec. Nov. Regni Veg.* 16: 378 (1920).
Hochstetter's butterfly orchid

SYNONYMS. *Habenaria longebracteata* Hochst. ex Seub.; *Limnorchis longebracteata* (Hochst. ex Seub.) Efimov.

DISTRIBUTION. Endemic to the Azores.

HABITAT. Grassy laurel scrub in the cloud zone; 950–1,000 m.

FLOWERING. June to August.

DISTINGUISHING FEATURES. 10–30 cm tall. Flowers whitish green, significantly larger than those of the other Azorean species: lateral sepals 7–10 mm long; petals 4–7 mm; lip 6.5–10 mm, reflexed; spur 8–11.5 mm. Column with the viscidia placed more than 2 mm apart.

LEFT TO RIGHT
Azores, São Jorge 16.06.2012 (30)
Azores, São Jorge 15.06.2012 (30)

Platanthera micrantha (Hochst. ex Seub.) Schltr., *Repert. Spec. Nov. Regni Veg.* 16: 378 (1920).
Small-flowered butterfly orchid
SYNONYM. *Habenaria micrantha* Hochst. ex Seub.
DISTRIBUTION. Endemic to the Azores, found on eight of the nine islands.
HABITAT. Laurel scrubland and adjacent grassland, on wet and acidic soil; 200 to 1,100 m.
FLOWERING. May and June.
DISTINGUISHING FEATURES. 15–50 cm tall. Plants with a dense spike of small pale green to cream flowers having deflexed lateral sepals and a porrect, upcurved, 3–6 mm long lip with a 6–9 mm long spur. Column with the viscidia placed less than 2 mm apart.

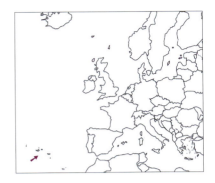

LEFT TO RIGHT
Azores, São Miguel 16.06.2001 (30)
Azores, São Miguel 16.05.1989 (5)

Platanthera pollostantha R.M.Bateman & M.Moura, *PeerJ* 1 (e218): 60 (2013).
Short-spurred butterfly orchid
DISTRIBUTION. Endemic to the Azores, found on all the islands.
HABITAT. Mainly in laurel scrub and grassland but also in rough pastures, oak woods and *Cryptomeria* plantations, on wet and acidic soil; 240 to 1,330 m.
FLOWERING. May and June.
DISTINGUISHING FEATURES. 10–45 cm. Similar to *P. micrantha* but differs in having horizontally spreading to recurved lateral sepals, a downwards-directed lip with recurved apex and a 2–5 mm long spur.

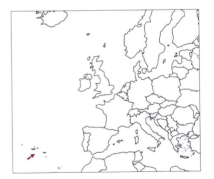

LEFT TO RIGHT
Azores, Pico 13.06.2012 (30)
Azores, Santa Maria 10.06.2012 (30)

20. GYMNADENIA R.Br. in Aiton, *Hortus Kew.* (Ed. 2) 5: 191 (1813).

A genus of 5 species in Europe, extending across temperate Asia. The tubers are broad, flat and divided. The 3–10 leaves are linear to narrowly lanceolate, folded and unspotted. The bracts are as long as or longer than the ovary. The mostly dense inflorescence has rose to pink or white, resupinate flowers with a slenderly spurred lip and freely exposed viscidia. Identification key on p. 425.

Gymnadenia frivaldii Hampe ex Griseb., *Spic. Fl. Rumel.* 2: 363 (1846). **Frivald's orchid**

SYNONYMS. *Pseudorchis frivaldii* (Hampe ex Griseb.) P.F. Hunt; *Leucorchis frivaldii* (Hampe ex Griseb.) Schltr.

DISTRIBUTION. From the S Carpathians across the Balkans to Greece.

HABITAT. In full sun in fens, seepages and wet meadows on siliceous soil; 1,000–2,300 m.

FLOWERING. June and July.

DISTINGUISHING FEATURES. 10–30 cm tall. Distinguished by its small habit and shortly conical, dense spike of pinkish white resupinate flowers with 3.5–4.5 mm long sepals and a broad, very obscurely trilobed lip with a slender, arching, 1.5–3 mm long spur. A local species but often found in sizeable colonies.

Gymnadenia densiflora (Wahlenb.) A.Dietr., *Allg. Gartenzeit.* 7: 170 (1839).
Dense-flowered fragrant orchid

SYNONYMS. *Orchis conopsea* var. *densiflora* Wahlenb.; *Gymnadenia conopsea* subsp. *densiflora* (Wahlenb.) K.Richt.

DISTRIBUTION. W and C Europe, across to the Balkans.

HABITAT. In seepages in chalk downland, and in base-rich fens; sea level to 2,800 m.

FLOWERING. June to August.

DISTINGUISHING FEATURES. 20–80 cm. A robust species that flowers 2 weeks to a month later than *G. conopsea* when they are sympatric. Flowers are usually darker rose-purple, with a flat lip with distinct shoulders, and have a distinctive carnation (*Dianthus*) scent.

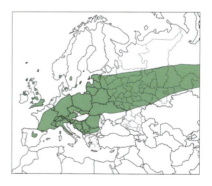

GYMNADENIA | 111

Greece, Kastoria
05.06.1987 (1)

Greece, C Macedonia
16.06.1998 (45)

Greece, C Macedonia
16.06.1998 (45)

England, Hampshire
07.07.2020 (25)

England, Hampshire
07.07.2020 (25)

Switzerland, Grisons
18.07.2002

GYMNADENIA

Gymnadenia borealis (Druce) R.M.Bateman, Pridgeon & M.W.Chase, *Lindleyana* 12 (2): 130 (1997). **Bog fragrant orchid**

SYNONYMS. *Habenaria gymnadenia* var. *borealis* Druce; *Gymnadenia conopsea* var. *borealis* (Druce) Soó; *G. conopsea* subsp. *borealis* (Druce) F.Rose

DISTRIBUTION. Britain, Ireland (very rare); possibly more widely distributed but verification of records from outside the British Isles is pending.

HABITAT. Acid grassland and marshy ground; up to 700 m.

FLOWERING. June to August.

DISTINGUISHING FEATURES. 10–20 cm tall. A smaller plant with a slenderer habit than *G. conopsea* and *G. densiflora*. Inflorescence with 20–30, pale lilac to deep pink, clove-scented flowers with small, flat and pointed lateral sepals and a small lip with 2 small side lobes, a longer mid-lobe and a shorter spur.

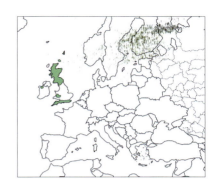

LEFT TO RIGHT
England, Cumbria 05.07.1996 (22)
Eire, County Clare 18.07.1995 (22)
Wales, Anglesey 24.06.1992 (22)

Gymnadenia conopsea (L.) R.Br. in Aiton, *Hortus Kew.* (Ed. 2) 5: 191 (1813). **Fragrant orchid**

SYNONYM. *Orchis conopsea* L.

DISTRIBUTION. Widespread and often common in Europe; also in Asia across to China, Korea and Japan and south to the Himalayas.

HABITAT. Usually in calcareous grassland and scrub; sea level to 2,800 m.

FLOWERING. May to August.

DISTINGUISHING FEATURES. 20–60 cm tall. Inflorescence usually lax, with many pink (to white or purple) flowers that are sweetly scented. Lip is distinctly trilobed with a slender, elongate, nectar-bearing spur.

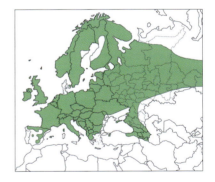

CLOCKWISE FROM TOP LEFT
Italy, South Tyrol 13.07.2007
Italy, South Tyrol 13.07.2007
Italy, Umbria 09.06.2016 (21)
Switzerland, Grisons 03.07.2006
Italy, Umbria 09.06.2016 (21)

Gymnadenia odoratissima (L.) Rich., *De Orchid. Eur.*: 35 (1817). **Short-spurred fragrant orchid**

SYNONYMS. *Orchis odoratissima* L., *Habenaria odoratissima* (L.) Franch.

DISTRIBUTION. Mainly from NE Spain to Germany, Austria and N Italy but with satellite ranges in Sweden, E Europe (to Belarus) and the Balkans.

HABITAT. In mountain meadows, semiarid grassland, wet meadows and fens, in alkaline soils, up to 2,800 m.

FLOWERING. June to August.

DISTINGUISHING FEATURES. 10–35 cm, or more, tall. Flowers smell of vanilla and have a 4–6 mm long spur that is shorter than that in *G. conopsea* (more than 10 mm long).

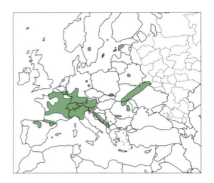

LEFT TO RIGHT
Switzerland, Grisons 19.07.2002
Switzerland, Grisons 22.07.2004
Switzerland, Grisons 12.07.2006
Switzerland, Grisons 16.07.2008
Switzerland, Baselland 22.06.2001
France, Aveyron 26.05.2007

21. NIGRITELLA Rich., *De Orchid. Eur.*: 26 (1817).

A genus endemic to Europe, comprising c. 10 species. The tubers are broad, flat and divided. The up to 10 leaves are linear to narrowly lanceolate, folded and unspotted. The bracts are as long as or longer than the ovary. The very dense, ovoid to hemispherical inflorescence has whitish to pink, scarlet or dark brown-red, non-resupinate flowers with freely exposed viscidia. The lip has a spur that is less than 1.7 mm long. The species are very difficult to distinguish from each other and a 20× hand lens is helpful in identifying them. In the 1st edition, we included *Nigritella* In a broadly defined *Gymnadenia*. The rationale for this change is provided in the introductory chapter on Taxonomy of European Orchids. Identification key on p. 426.

Nigritella gabasiana Teppner & E.Klein, *Phyton (Horn)* 33: 182 (1993). **Spanish vanilla orchid**

SYNONYM. *Gymnadenia gabasiana* (Teppner & E.Klein) Teppner & E.Klein

DISTRIBUTION. The Pyrenees and the Cantabrian mountains.

HABITAT. Grassland and meadows, mainly on calcareous soil; 1,500–2,100 m.

FLOWERING. June to August.

DISTINGUISHING FEATURES. Inflorescence subconical to hemispherical at peak flowering. Bracts papillate except at the tip. Flower buds blackish-red to purplish-red or yellow; flowers not losing colour with age, unscented to faintly aromatic. Lip sides markedly incurved in the basal third to half of the lip.

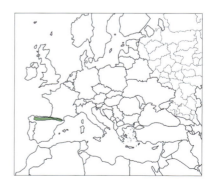

BOTTOM ROW, LEFT TO RIGHT
France, Pyrenees 21.06.2012 (62)
France, Pyrenees 21.06.2012 (62)
France, Pyrenees 21.06.2012 (62)

Nigritella lithopolitanica Ravnik, *Acta Bot. Croat.* 37: 226 (1978). **Yugoslavian vanilla orchid**

SYNONYM. *Gymnadenia lithopolitanica* (Ravnik) Teppner & E.Klein

DISTRIBUTION. Endemic to the SE Alps in Austria and Slovenia.

HABITAT. Grassland on calcareous soil; 1,500–2,000 m.

FLOWERING. June and July.

DISTINGUISHING FEATURES. Flower buds pink to rose-coloured. Lip sides markedly incurved in the basal third to half of the lip, forming a tube; basal part of the lip (in its natural conformation) narrower than the apical part. Rostellum not protruding between the anther thecae.

LEFT TO RIGHT
Austria, Hochobir 25.06.2008 (62)
Austria, Hochobir 25.06.2008 (62)
Austria, Kärnten 23.06.2003 (3)

Nigritella rhellicani Teppner & E.Klein, *Phyton (Horn)* 31: 7 (1990). **Alpine vanilla orchid**

SYNONYMS. *Gymnadenia rhellicani* (Teppner & E. Klein) Teppner & E.Klein; ?*Gymnadenia carpatica* (Zapal.) Teppner & E.Klein

DISTRIBUTION. The Alps, the Apennines, the Balkans (to N Greece) and possibly the E Carpathians.

HABITAT. Grassland and meadows, mainly on calcareous soil; 1,000–2,800 m.

FLOWERING. June to August.

DISTINGUISHING FEATURES. Lower bracts minutely papillose-dentate. Flower buds usually blackish-red (but variable); flowers not losing colour with age, vanilla-scented. Lip sides only at the very base of the lip markedly incurved; spur 1–1.6 mm long.

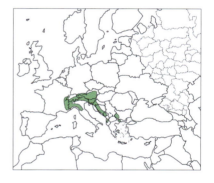

LEFT TO RIGHT
Italy, South Tyrol 07.07.2010
Switzerland, Grisons 15.07.1995
Switzerland, Grisons 16.07.2001
Switzerland, Grisons 16.07.2008
Switzerland, Grisons 16.07.2008

Nigritella nigra (L.) Rchb.f. in Rchb., *Icon. Fl. Germ. Helv.* 13-14: 102 (1851). **Vanilla orchid**
SYNONYMS. *Satyrium nigrum* L.

a. subsp. *nigra*
DISTRIBUTION. C Sweden and C and N Norway.
HABITAT. Grassland and hay meadows on calcareous soil; near sea level to 1,300 m.
FLOWERING. June to early August.
DISTINGUISHING FEATURES. Lower bracts (sub)entire. Flower buds dark (brownish-)red; flowers not losing colour with age, vanilla-scented. Lateral sepals straight to slightly recurved. Lip sides only at the very base of the lip markedly incurved; spur 0.7–1 mm long.

b. *Nigritella nigra* subsp. *austriaca* Teppner & E.Klein, *Phyton (Horn)* 31: 17 (1990).
Austrian vanilla orchid
SYNONYMS. *Gymnadenia nigra* subsp. *austriaca* (Teppner & E.Klein) Teppner & E.Klein; *N. austriaca* (Teppner & E.Klein) P.Delforge; *N. nigra* subsp. *gallica* E.Breiner & R.Breiner; *N. nigra* subsp. *iberica* Teppner & E.Klein; *Gymnadenia austriaca* (Teppner & E.Klein) P.Delforge; *G. austriaca* var. *gallica* (E. & R.Breiner) P.Delforge
DISTRIBUTION. From the Pyrenees to the Alps and in the Carpathians.
HABITAT. Grassland and meadows, mainly on calcareous soil; (1,100–)1,400–2,400 m.
FLOWERING. June to August.
DISTINGUISHING FEATURES. Lower bracts (sub)entire. Flower buds dark red to blackish-red; flowers not losing colour with age, chocolate-scented. Lip sides only at the very base of the lip markedly incurved; spur 1–1.3 mm long.

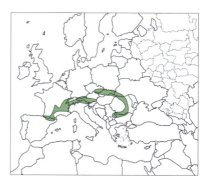

Norway, Sør-Trøndelag
07.07.2004 (45)

Norway, Sør-Trøndelag
10.07.2004 (45)

Norway, Sør-Trøndelag
07.07.2004 (45)

Italy, S Tyrol
07.07.2010

Italy, S Tyrol
07.07.2010

Italy, S Tyrol
07.07.2010

NIGRITELLA

France, Haute Alpes
10.07.2001 (50)

France, Haute Alpes
10.07.2001 (50)

France, Dep. Mayenne
13.07.2003 (50)

Italy, South Tyrol
15.07.2002

Austria, Steiermark
26.04.2004 (3)

Nigritella corneliana (Beauverd) Gölz &
H.R.Reinhard, *Jahresber. Naturwiss. Vereins
Wuppertal* 39: 39 (1986). **Cornel's vanilla orchid**

SYNONYMS. *Nigritella nigra* subsp. *corneliana*
Beauverd; *N. lithopolitanica* subsp. *corneliana*
(Beauverd) Teppner & E.Klein; *N. corneliana* subsp.
bourneriasii E. & R.Breiner; *Gymnadenia corneliana*
(Beauvaerd) Teppner & E.Klein

DISTRIBUTION. Endemic to the SW Alps of France and Italy.

HABITAT. Grassland on calcareous soil; 1,500–2,500 m.

FLOWERING. June to August.

DISTINGUISHING FEATURES. Stem slender (upper part about as wide as base of the lowermost bracts). Lower bracts minutely papillose-dentate with 0.02–0.03 mm long teeth. Flower buds red; flowers losing colour with age. Lip sides markedly incurved in the basal half to third of the lip.

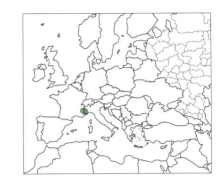

Nigritella widderi Teppner & E.Klein, *Phyton (Horn)* 25: 318 (1985). **Widder's vanilla orchid**

SYNONYM. *Gymnadenia widderi* (Teppner & E.Klein) Teppner & E.Klein

DISTRIBUTION. NE Alps to C Italy.

HABITAT. Grassland on calcareous soils; 1,500–2,200 m.

FLOWERING. June to August.

DISTINGUISHING FEATURES. Flower buds pink to rose-coloured. Lip sides markedly incurved in the basal half to third of the lip; basal part of the lip (in its natural conformation) distinctly wider than the apical part. Rostellum protruding between the anther thecae.

NOTE. *Nigritella archiducis-joannis* Teppner & E.Klein (synonym *Gymnadenia archiducis-joannis* (Teppner & E.Klein) Teppner & E.Klein) from the Austrian Alps might deserve taxonomic recognition. It differs from *N. widderi* in having flowers that do not open fully, but genetic separation of the two taxa seems obscure.

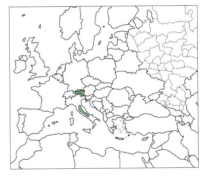

Nigritella buschmanniae Teppner & Ster, *Phyton (Horn)* 36: 278 (1996). **Buschmann's vanilla orchid**
SYNONYM. *Gymnadenia buschmanniae* (Teppner & Ster) Teppner & E.Klein
DISTRIBUTION. Endemic to NE Italy.
HABITAT. Grassland on calcareous soil; 2,300–2,400 m.
FLOWERING. July and August.
DISTINGUISHING FEATURES. Stem stout (upper part at least 1.5 times as wide as the base of the lowermost bracts). Lower bracts minutely papillose-dentate with c. 0.1 mm long teeth. Flower buds red; flowers losing colour with age. Lip sides markedly incurved in the basal third to half of the lip.

Nigritella miniata (Crantz) Janch., *Phyton (Horn)* 8: 232 (1959). **Red vanilla orchid**
SYNONYM. *Orchis miniata* Crantz; *Nigritella miniata* (Crantz) Janch.; *Gymnadenia rubra* Wettst.; *Nigritella rubra* (Wettst.) K.Richt.; *Gymnadenia miniata* (Crantz) Hayek; *G. dolomitensis* Teppner & E.Klein; *Nigritella bicolor* W.Foelsche; *N. hygrophila* W.Foelsche & Heidtke
DISTRIBUTION. C and SE Alps and the S Carpathians.
HABITAT. Grassland and meadows on calcareous soil; (980–)1,400–2,600 m.
FLOWERING. June to August.
DISTINGUISHING FEATURES. Inflorescence ovoid at peak flowering. Flower buds dark ruby-red; flowers not losing colour with age, vanilla-scented. Lip sides markedly incurved in the basal half to third of the lip.
NOTE. A population at Trenchtling in Austria is characterized by remarkably small perianths. It has been described as *N. minor* W.Foelsche & Zernig and might deserve taxonomic recognition.

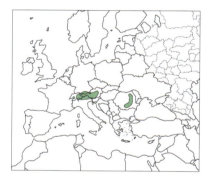

Switzerland, Grisons
10.07.2009

Italy, Brenta Dolomites
15.07.1998 (3)

Italy, P. so del Griste
16.07.2005 (62)

Italy, P. so del Griste
16.07.2005 (62)

Italy, Veneto
20.06.1990 (45)

Switzerland, Ticino
10.07.1988 (11)

Nigritella stiriaca (Rech.) Teppner & E.Klein, *Phyton (Horn)* 25: 159 (1985)

SYNONYMS. *Gymnadenia rubra* var. *stiriaca* Rech.; *G. stiriaca* (Rech.) Teppner & E.Klein

DISTRIBUTION. Only known by certainty from a small area in the north-eastern limestone Alps of Austria.

HABITAT. Grassland on calcareous soils; 1,250–2,000 m.

FLOWERING. June to July.

DISTINGUISHING FEATURES. Similar to *N. miniata*, but differs in its perianth being rose-coloured (proximally) to whitish (distally).

ALL IMAGES
Austria, Schafberg, St Wolfgang 24.06.2009 (62)

22. GYMNIGRITELLA E.G.Camus, *J. Bot. (Morot)* 6: 484 (1892).

Whereas any fortuitous hybrid between two species belonging to *Gymnadenia* and *Nigritella*, respectively, can formally be recognized by a binary name under *Gymnigritella*, this genus only contains one species forming self-reproducing populations. That is the tetraploid, apomictic, Swedish endemic, *Gymnigritella runei* which appears to have evolved through hybridization between *Gymnadenia conopsea* and *Nigritella nigra* subsp. *nigra*. Being morphologically highly similar to *Nigritella*, it mainly differs in having wine-red flowers and a slightly longer spur (> 1.8 mm). This was treated as *Gymnadenia runei* in the first edition.

Gymnigritella runei Teppner & E.Klein, *Phyton (Horn)* 29: 163 (1989). **Rune's vanilla orchid**

SYNONYMS. *Gymnadenia runei* (Teppner & E.Klein) Ericsson; *Nigritella runei* (Teppner & E.Klein) Kreutz

DISTRIBUTION. Endemic to Swedish Lapland.

HABITAT. Grassland, *Dryas octopetala* heaths and marshes on calcareous soil; around 800 m.

FLOWERING. July.

DISTINGUISHING FEATURES. Flower buds deep wine-red; flowers not losing colour with age, faintly sweet-scented. Lateral sepals distinctly recurved. Lip sides only at the very base of the lip markedly incurved; spur 1.8–2.3 mm long.

LEFT TO RIGHT
Sweden, Åsele Lappmark 17.07.2005 (45)
Sweden, Åsele Lappmark 17.07.2005 (45)
Sweden, Åsele Lappmark 17.07.2005 (45)

NIGRITELLA – GYMNIGRITELLA 125

23. COELOGLOSSUM Hartm., *Handb. Skand. Fl.*: 323 (1920).

A monotypic genus distributed in temperate to subarctic regions in the Northern Hemisphere, and with scattered montane occurrences in neighbouring subtropical regions. The rationale for accepting *Coeloglossum* as distinct from *Dactylorhiza* is given in the introductory chapter on Taxonomy of European Orchids.

Coeloglossum viride (L.) Hartm., *Handb. Skand. Fl.*: 329 (1820). **Frog orchid**

SYNONYMS. *Satyrium viride* L.; *Dactylorhiza viridis* (L.) R.M.Bateman, Pridgeon & M.W.Chase

DISTRIBUTION. Temperate to subarctic regions across the northern hemisphere, and with scattered montane occurrences in neighbouring subtropical regions.

HABITAT. In full sun to light shade in heaths, meadows, grassland, fens, scrub and open woodland, on acidic to alkaline soils; sea level to 4,000 m (in Europe to 2,970 m).

FLOWERING. May to August.

DISTINGUISHING FEATURES. 5–30(–40) cm tall. Easily recognizable by the (yellow-)green to (red-)brown flowers in which the sepals and petals form a hood, by the elongate, distally 3-dentate lip that bears a short sac-like spur and by the column producing 2 widely separated bursicles.

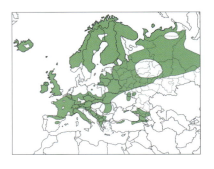

BELOW, LEFT TO RIGHT
Switzerland, Grisons 11.07.2013
Finland, Lapland 11.06.2014 (40)

OPPOSITE, LEFT TO RIGHT FROM TOP
Switzerland, Grisons 11.07.2013
Switzerland, Grisons 01.07.2018
Italy, Lazio 20.06.2004 (21)
Switzerland, Grisons 03.07.1977 (11)
Switzerland, Grisons 20.06.1999
Italy, South Tyrol 15.07.2009
Switzerland, Grisons 17.07.2006
Switzerland, Grisons 14.07.2007
Switzerland, Grisons 08.07.2001

24. DACTYLORHIZA Neck. ex Nevski, *Trudy Bot. Inst. Akad. Nauk S.S.S.R., Ser. 1, Fl. Sist. Vyssh. Rast.* 4: 332 (1937), nom. cons.

SYNONYM. *Dactylorchis* (Klinge) Verm.

DISTRIBUTION. Temperate to subarctic regions in the Northern Hemisphere, and with scattered montane occurrences in neighbouring subtropical regions.

Twelve species in our region. Largely distinguished by the herbaceous bracts that are not pressed against the ovaries, by the spreading to reflexed lateral sepals, by the lip having a conical to cylindrical spur and by the column producing a single bursicle.

Dactylorhiza is a difficult genus with a series of morphologically similar taxa. Their taxonomy is complicated by their propensity to hybridise where two or more taxa are sympatric, sometimes undergoing polyploidisation (doubling of the chromosome number) in the process. Many classical morphological studies have failed to clarify the relationships within this genus, but a better understanding has been obtained by a combination of DNA, chromosomal and morphometric analyses. In our opinion, morphologically distinct allopolyploid taxa that have identical chromosome numbers and that share the same species as parents should only be separated from each other at infraspecific rank. Hence, we recognise the following allopolyploid species: *D. insularis* (combines a diploid genome from *D. romana* s.l. with a haploid genome from *D. sambucina*), *D. cantabrica* (combines diploid genomes from *D. romana* s.l. and *D. sambucina*), *D. armeniaca* (combines diploid genomes from *D. euxina* and *D. incarnata* s.l.), *D. majalis* (combines diploid genomes from *D. incarnata* s.l. and *D. maculata* s.l.) and *D. urvilleana* (combines diploid genomes from *D. euxina* and *D. maculata* s.l.).

Hybrid swarms are not uncommon because species often flower more or less at the same time in the same habitat. Such widespread hybridisation has clearly been one of several factors underlying the description of a greatly exaggerated number of species and infraspecific taxa in *Dactylorhiza*. Other factors include phenotypic plasticity and genetically determined polymorphism. Studies that are based on molecular data and/or appropriately collected morphometric data support a much smaller number of taxa than that recognised by most authors.

For identification, it is usually important to check whether the stem – just below the inflorescence – is squeezable (i.e. hollow) or non-squeezable (i.e. solid or very nearly so). Another important character is whether the bracts are quite entire or minutely crenate or serrate because of enlarged and protruding marginal cells; this can be checked using a strong (20×) hand lens and natural backlight. Floral traits should always be scored from fully opened flowers in the mid-portion of the spike (unless otherwise stated). Finally, it should be noted that, in the accounts below, distinction is made between cataphylls, sheathing leaves and non-sheathing leaves. Identification key on p. 426.

Dactylorhiza iberica (M.Bieb. ex Willd.) Soó, *Nom. Nov. Gen. Dactylorhiza*: 3 (1962).
Crimean marsh-orchid

SYNONYM. *Orchis iberica* M.Bieb. ex Willd.
DISTRIBUTION. From Greece across Turkey to the Crimea, the Caucasus, NW Iran, Cyprus and Lebanon.
HABITAT. Spring-fed swamps, wet meadows, by lakes and rivers, mainly on calcareous soil; 360–2,600 m.
FLOWERING. May to August.
DISTINGUISHING FEATURES. 15–50(–60) cm tall, forming underground stolons. Leaves grass-like, suberect, linear-lanceolate, unspotted. Flowers rose to mauve, less frequently white or purple. Lateral sepals connivent with petals and dorsal sepal. Labellum distally 3-lobed, cuneate at base, marked with spots and streaks; spur thinner and shorter than ovary.

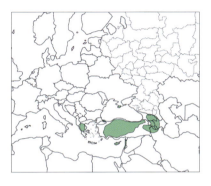

FAR LEFT TO RIGHT
Greece, Peloponnese 30.07.1990 (1)
Georgia, Darial gorge 14.07.2014 (24)
Georgia, Darial gorge 14.07.2014 (24)
Azerbaijan, Baku 03.06.1997 (1)

Dactylorhiza sambucina (L.) Soó, *Nom. Nov. Gen.* Dactylorhiza: 3 (1962). **Elder-flowered orchid**

SYNONYM. *Orchis sambucina* L.

DISTRIBUTION. S, C and E Europe to S Fennoscandia, and with an isolated occurrence in the Lesser Caucasus.

HABITAT. In full sun to light shade on grassland and in mountain meadows and open conifer forest, on moderately dry, fairly acidic to alkaline soil; sea level to 2,600 m.

FLOWERING. Late April to early July.

DISTINGUISHING FEATURES. 9–25(–40) cm tall. Sheathing leaves alternate, unspotted; angle between stem and longest leaf > 55°. Flowers yellow, salmon-pink or crimson-mauve to crimson-purple (with yellow lip base). Lateral sepals reflexed and erect or nearly so. Lip straight (i.e. not recurved), marked with dots or dashes that usually extend beyond its proximal third, 3-lobed with emarginate to truncate (rarely rounded) mid-lobe; spur straight to downcurved, > 11 mm long, > 1.4 times as long as the lip lamina, thicker than the ovary.

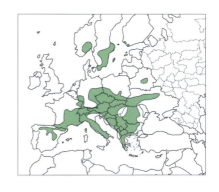

BELOW, LEFT TO RIGHT
France, Aveyron 20.05.2004
France, Aveyron 15.05.2009

OPPOSITE, CLOCKWISE FROM TOP LEFT
France, Aveyron 25.05.2009
France, Aveyron 20.05.2009
France, Aveyron 20.05.2004
France, Aveyron 25.05.2016
France, Haute Provence 08.05.2010
France, Aveyron 21.05.1996

Dactylorhiza romana (Sebast.) Soó, *Nom. Nov. Gen.* Dactylorhiza: 3 (1962).
SYNONYM. *Orchis romana* Sebast.

a. subsp. *romana* Roman orchid
SYNONYM. *Dactylorhiza libanotica* (Mouterde) Aver.
DISTRIBUTION. S Italy to the Crimea, E Anatolia and N Israel.
HABITAT. In light to medium shade in open deciduous and coniferous forest or woodland and garigue, on moderately dry, alkaline to slightly acidic soil; sea level to 2,000 m.
FLOWERING. March to May (to early June).
DISTINGUISHING FEATURES. Usually < 23 cm tall. Sheathing leaves forming a basal rosette, unspotted. Non-sheathing leaves below inflorescence < 4 in number. Flowers cream to yellow or rose to purple (often with a yellow lip base), very rarely white or bicoloured. Lateral sepals reflexed and erect or nearly so. Lip devoid of markings, 3-lobed with emarginate to truncate (rarely rounded) mid-lobe; spur upcurved, > 12 mm long, not distinctly thicker than the ovary, vertical diameter of entrance > 2.5 mm.

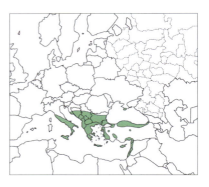

BELOW, LEFT TO RIGHT
Italy, Tuscany 12.04.2006
Italy, Tuscany 12.04.2006
Greece, Pelion 06.04.2022 (25)

OPPOSITE, LEFT TO RIGHT FROM TOP
Italy, Tuscany 13.04.1996
Cyprus 18.03.1996
Italy, Tuscany 10.04.2001
Cyprus 18.03.2007
Italy, Tuscany 12.04.2006
Cyprus 17.03.2009
Italy, Tuscany 12.04.2006
Italy, Tuscany 25.04.1996
Italy, Tuscany 25.04.1996

b. *Dactylorhiza romana* subsp. *georgica*

(Klinge) Renz & Taubenheim, *Notes Roy. Bot. Gard. Edinburgh* 41: 271 (1983).

SYNONYMS. *Orchis mediterranea* subsp. *georgica* Klinge; *Dactylorhiza flavescens* (K.Koch) Holub.

DISTRIBUTION. From E Anatolia to the Caucasus, Transcaucasia, N Iran and S Turkmenistan.

HABITAT. Open forest, grassland and fairly dry mountain meadows, on alkaline to slightly acidic soils; (550–)900–2,200 m.

FLOWERING. Late March to early June.

DISTINGUISHING FEATURES. Differs from the other two subspecies in having a narrower spur entrance (vertical diameter < 2.5 mm). In contrast to the consistently yellow-flowered subsp. *guimaraesii*, individuals of subsp. *georgica* have either cream to yellow or rose to purple flowers.

c. *Dactylorhiza romana* subsp. *guimaraesii*

(E.G.Camus) H.A.Pedersen, *Bot. J. Linn. Soc.* 152: 426 (2006).

SYNONYMS. *Orchis romana* var. *guimaraesii* E.G.Camus; *Dactylorhiza guimaraesii* (E.G.Camus) P.Delforge

DISTRIBUTION. Iberian Peninsula, N Morocco and N Algeria.

HABITAT. In full sun to light shade on grassland and in wooded meadows and open forest, on neutral to acidic soils; 700–2,000 m.

FLOWERING. March to early June.

DISTINGUISHING FEATURES. Differs from the other two subspecies by having the following combination of features: usually > 23 cm tall; non-sheathing leaves below inflorescence > 4 in number; flowers yellow; spur < 12 mm long, vertical diameter of entrance > 2.5 mm.

Azerbaijan, Baku
22.04.2000 (1)

Turkey, Artvin
17.05.2000 (45)

Turkey, Artvin
17.05.2000 (45)

Spain, Navarre
11.05.2014 (45)

Spain, Navarre
11.05.2014 (45)

Spain, Navarre
11.05.2014 (45)

Dactylorhiza insularis (Sommier) Landwehr, *Orchidee (Hamburg)* 20: 128 (1969).

SYNONYM. *Orchis insularis* Sommier

DISTRIBUTION. W Mediterranean to WC Italy, Corsica, Sardinia and N Morocco.

HABITAT. Open pine, oak and chestnut forest and sparse mountain meadows, on acidic soil; sea level to 1,500(–2,000) m.

FLOWERING. April to June.

DISTINGUISHING FEATURES. 10–35(–50) cm tall. Most, if not all, sheathing leaves forming a basal rosette, unspotted. Flowers yellow. Lateral sepals reflexed and erect or nearly so. Lip provided with (often few and weak) red markings that do not extend beyond its proximal third, 3-lobed with emarginate to truncate (rarely rounded) mid-lobe; spur straight to downcurved, < 11 mm long, < 1.4 times as long as the lip lamina, not distinctly thicker than the ovary.

ALL IMAGES
Italy, Tuscany 18.04.1998

Dactylorhiza cantabrica H.A.Pedersen, *Bot. J. Linn. Soc.* 152: 428 (2006). **Galician orchid**

DISTRIBUTION. Western part of the Cantabrian Mountains (Spain).

HABITAT. Fairly dry montane grassland; 1,200–1,500 m.

FLOWERING. May to early June.

DISTINGUISHING FEATURES. 10–20 cm tall. Similar to yellow-flowered individuals of *D. sambucina*, but differs in the angle between the stem and longest leaf being < 55° and in the spur being < 11 mm long and < 1.4 times as long as the lip lamina. Besides, the lip is usually more or less recurved.

BOTH IMAGES
Spain, Galicia 28.04.2001 (45)

Dactylorhiza euxina (Nevski) Czerep., *Sosud. Rast. SSSR*: 308 (1981).

SYNONYM. *Orchis euxina* Nevski.

DISTRIBUTION. NE Turkey and the Caucasus.

HABITAT. In full sun to light shade in wet meadows, fens and short grassland, on slightly acidic to slightly alkaline and nutrient-rich soil; (500–)850–2,900 m.

FLOWERING. Mainly from June until early August, but already in May at the few lowland sites.

DISTINGUISHING FEATURES. 5–30(–50) cm tall. Stem squeezable, upper two-thirds (inflorescence not included) bearing at least as many nodes of sheathing leaves as lowermost third. Leaves usually with numerous small purplish-brown spots and streaks on both surfaces, less frequently unspotted. Bracts minutely crenate or serrate (observe the bract margin against the light using a 20× hand lens!). Flowers mauve to purple with more or less confluent markings on the lip (rarely white and without lip markings). Lip with erose lateral margins; spur conical in the lower view.

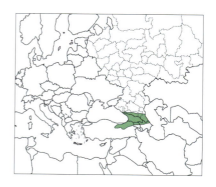

ALL IMAGES
Turkey, Rize, 20.05.2000 (45)

Dactylorhiza incarnata (L.) Soó, *Nom. Nov. Gen. Dactylorhiza*: 3 (1962).

SYNONYM. *Orchis incarnata* L.

a.1. subsp. *incarnata* var. *incarnata* Early marsh-orchid

SYNONYMS. *Dactylorhiza incarnata* var. *drudei* (M.Schulze) Soó; *D. incarnata* var. *macrophylla* (Schur) Soó; *D. incarnata* subsp. *pulchella* (Druce) Soó; *D. incarnata* var. *hyphaematodes* (Neuman) Løjtnant; *D. pulchella* (Druce) Aver.; *D. incarnata* subsp. *baumgartneriana* B. & H.Baumann, R.Lorenz & Ruedi Peter; *D. incarnata* var. *baumgartneriana* (B. & H.Baumann, R.Lorenz & Ruedi Peter) P.Delforge; *D. incarnata* var. *haussknechtii* (Klinge) Buttler; *D. incarnata* subsp. *jugicrucis* Akhalk., R.Lorenz & Mosul.; *D. incarnata* var. *jugicrucis* (Akhalk., R.Lorenz & Mosul.) P.Delforge

DISTRIBUTION. Widespread across Europe and temperate Asia to Lake Baikal, Mongolia and NW China.

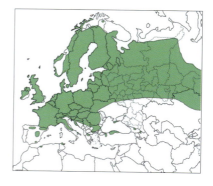

DACTYLORHIZA

Russia, Pskov Region 2
26.06.2015 (46)

Switzerland, Grison
06.07.2019

Switzerland, Grison
12.07.2011

HABITAT. Spring-fed swamps, fens and wet meadows on calcareous, neutral or acidic soil; sea level to 2,400 m.

FLOWERING. May to early August.

DISTINGUISHING FEATURES. Plant slender (longest leaf usually < 0.6 times and inflorescence < 0.3 times as long as the plant is tall). Stem squeezable; upper two-thirds (inflorescence not included) bearing fewer nodes of sheathing leaves than lowermost third. Leaves unspotted or marked with tiny spots (all < 1 mm in diameter) on one or both surfaces; sheathing leaves spreading to suberect (rarely erect). Bracts quite entire (observe the bract margin against the light using a 20× hand lens!). Flowers usually rose to purple with lip markings, occasionally white to cream and/or without lip markings. Lateral sepals reflexed and erect. Petals devoid of markings. Lip shallowly 3-lobed with acuminate to rounded (rarely truncate) mid-lobe and entire or obscurely notched side lobes, usually < 9 mm wide when flattened; spur thicker than the ovary; lip of the lowermost flower, when flattened, less than twice as wide as the stem just below the inflorescence (except sometimes in dwarf individuals).

France
12.06.1999 (40)

a.2. *Dactylorhiza incarnata* subsp. *incarnata* var. *cruenta* (O.F.Müll.) Hyl., *Nord. Kärlväxtfl.* 2: 387 (1966).

SYNONYMS. *Orchis cruenta* O.F.Müll.; *Dactylorhiza cruenta* (O.F.Müll.) Soó; *D. incarnata* subsp. *cruenta* (O.F.Müll.) P.D.Sell.

DISTRIBUTION. Temperate Europe and across temperate Asia to Lake Baikal, Mongolia and NW China; in Europe, much less common than var. *incarnata*.

HABITAT. In full sun in coastal fens, flat moors and alpine meadows, on calcareous soils; sea level to 2,500 m.

FLOWERING. June to early August.

DISTINGUISHING FEATURES. Differs from the other four subspecies in having the following combination of features: leaves spotted purple-brown on both surfaces with at least some spots > 1 mm in diameter, rarely unspotted; bracts quite entire; flowers (purple-)mauve; petals marked with purple dashes and/or circles; lip entire to shallowly 3-lobed.

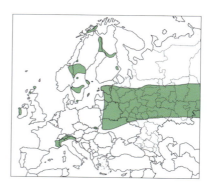

CLOCKWISE FROM TOP LEFT
var. *cruenta*
Switzerland, Grisons 06.07.2019
Italy, South Tyrol 06.07.2019
Switzerland, Grisons 15.07.2003
Italy, South Tyrol 12.07.2008
Switzerland, Grisons 15.07.2003

DACTYLORHIZA | 141

Switzerland, Aargau
02.07.1979 (8)

Switzerland, Grisons
23.07.1997

Switzerland, Grison
05.07.2013

a.3. *Dactylorhiza incarnata* subsp. *incarnata* var. *ochroleuca* (Boll) Hyl., *Nord. Kärlväxtfl.* 2: 387 (1966).

SYNONYMS. *Orchis incarnata* var. *ochroleuca* Wüstnei ex Boll; *Dactylorhiza ochroleuca* (Wüstnei ex Boll) Holub

DISTRIBUTION. E and C Europe to East Anglia in the west and Estonia in the northeast.

HABITAT. In full sunlight in calcareous fens and meadows; sea level to 900 m.

FLOWERING. Late May to early July.

DISTINGUISHING FEATURES. Differs from var. *incarnata* in the combination of (very nearly) erect leaves, cream to yellow flowers devoid of markings and the lip in most cases being moderately 3-lobed with distinctly notched side lobes.

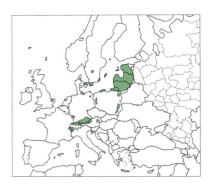

LEFT TO RIGHT, TOP ROW
Denmark, Sjælland 20.06.2002 (45)
Estonia, Saaremaa 24.06.2001 (45)
Estonia, Saaremaa 24.06.2001 (45)

b. *Dactylorhiza incarnata* subsp. *cilicica* (Klinge) H. Sund., *Europ. Medit. Orchid.*, ed. 2: 45 (1975).

SYNONYMS. *Orchis orientalis* subsp. *cilicica* Klinge; *Dactylorhiza umbrosa* (Kar. & Kir.) Nevski; *D. osmanica* (Klinge) Soó; *D. osmanica* subsp. *anatolica* (E.Nelson) Eccarius; *D. incarnata* var. *kotschyi* (Rchb.f.) H.A.Pedersen, P.J.Cribb & Rolf Kühn

DISTRIBUTION. From C Turkey to Lake Baikal in Russia.

HABITAT. In full sun in wet grassland, marshes, open swamp forest and by streams and lakes, on alkaline to slightly acidic soils; 500–3,800 m.

FLOWERING. May to July.

DISTINGUISHING FEATURES. Differs from the other two subspecies in its minutely crenate to serrate bracts (observe the bract margin against the light using a 20× hand lens!) and the generally larger flowers (lip usually > 9 mm wide when flattened).

NOTE. In the first edition of this field guide (Kühn et al. 2019), we treated this taxon at variety level under the name *D. incarnata* var. *kotschyi*. However, we now consider it sufficiently distinct (genetically and geographically) to be recognized as a separate subspecies (at which level the epithet *cilicica* has priority).

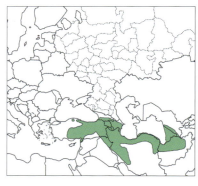

LEFT TO RIGHT, BOTTOM ROW
Turkey, Erzincan 26.05.2000 (45)
Turkey, Gümüshane 27.05.2000 (45)
Turkey, Erzincan 26.05.2000 (45)

c. *Dactylorhiza incarnata* subsp. *coccinea*

(Pugsley) Soó, *Nom. Nov. Gen. Dactylorhiza*: 4 (1962).

SYNONYMS. *Orchis latifolia* var. *coccinea* Pugsley; *Dactylorhiza incarnata* var. *lobelii* (Verm.) Soó; *D. incarnata* var. *dunensis* (Druce) Hyl.; *D. coccinea* (Druce) Aver.; *D. incarnata* subsp. *lobelii* (Verm.) H.A.Pedersen

DISTRIBUTION. Ireland, Great Britain, SW Norway, Denmark and The Netherlands.

HABITAT. In full sun in damp or wet meadows, fens and dune slacks, usually near the coast; sea level to 200 m.

FLOWERING. June to July.

DISTINGUISHING FEATURES. Differs from the other two subspecies in having the following combination of features: plant condensed (longest leaf usually > 0.6 times and inflorescence > 0.3 times as long as the plant is tall); leaves unspotted; bracts quite entire (observe the bract margin against the light using a 20× hand lens!); flowers crimson, rose or mauve; lip usually < 9 mm wide when flattened.

NOTE. In the first edition of this field guide (Kühn *et al.* 2019), we treated this taxon at variety level under the name *D. incarnata* var. *dunensis*. However, we now consider it sufficiently distinct (genetically and ecologically) to be recognized as a separate subspecies (at which level the epithet *coccinea* has priority).

LEFT TO RIGHT
Wales, Gwynedd 14.06.1999 (45)
The Netherlands, S Holland 07.06.2000 (45)
Wales, Gwynedd 14.06.1999 (45)

Dactylorhiza armeniaca Hedrén, *Pl. Syst. Evol.* 229: 42 (2001). **Armenian marsh-orchid**

SYNONYM. *Dactylorhiza euxina* subsp. *armeniaca* (Hedrén) Kreutz

DISTRIBUTION. NE Turkey to C Georgia.

HABITAT. In full sun to light shade in wet meadows (often with seepage water) and stream banks on slightly alkaline soils; 1,300–1,800 m.

FLOWERING. May to early July.

DISTINGUISHING FEATURES. (25–)30–65 cm, or rarely more, tall. Stem squeezable, upper two-thirds (inflorescence not included) bearing at least as many nodes of sheathing leaves as lowermost third. Leaves unspotted. Flowers mauve with distinct to more or less confluent markings on lip. Lip with (sub)entire lateral margins; spur (conical-)cylindrical in lower view.

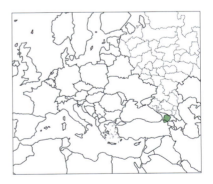

LEFT BOTH IMAGES
Georgia 30.05.2007 (27)

Dactylorhiza foliosa (Rchb.f.) Soó, *Nom. Nov. Gen.* Dactylorhiza: 7 (1962). **Madeiran orchid**
SYNONYM. *Orchis foliosa* Sol. ex Lowe
DISTRIBUTION. Endemic to Madeira.
HABITAT. Grows on slightly acidic soil on moist slopes and along levadas (irrigation canals); mainly found in humid and shady montane laurel forest on the north side of the island; 400–1,200(–1,500) m.
FLOWERING. May to July.
DISTINGUISHING FEATURES. 30–70 cm tall. Stem non-squeezable. Leaves green, unspotted. Flowers mauve to pink. Lateral sepals slightly spreading. Lip with obscure markings; spur thinner than the ovary.

LEFT TO RIGHT
Madeira 19.05.1995 (16)
Madeira 26.05.2006 (16)

Dactylorhiza maculata (L.) Soó, *Nom. Nov. Gen.* Dactylorhiza: 7 (1962).
SYNONYM. *Orchis maculata* L.

a. subsp. *maculata*
Heath spotted-orchid
SYNONYMS. *Dactylorhiza maculata* subsp. *islandica* (Á.Löve & D.Löve) Soó; *D. maculata* subsp. *transsilvanica* (Schur) Soó; *D. maculata* subsp. *caramulensis* Verm.; *D. maculata* var. *podesta* Landwehr; *D. maculata* subsp. *sudetica* (Poech ex Rchb.f.) Vöth; *D. ericetorum* (E.F.Linton) Aver.; *D. sudetica* (Poech ex Rchb.f.) Aver.; *D. islandica* (Á.Löve & D.Löve) Aver.; *D. kolaensis* (Montell) Aver.; *D. caramulensis* (Verm.) D.Tyteca; *D. savogiensis* D.Tyteca & Gathoye; *D. maculata* var. *elodes* (Griseb.) Aver.; *D. maculata* subsp. *podesta* (Landwehr) Kreutz; *D. maculata* var. *transsilvanica* (Schur) P.Delforge; *D. maculata* subsp. *savogiensis* (D.Tyteca & Gathoye) Kreutz; *D. maculata* subsp. *pyrenaica* Kreutz; *D. maculata* var. *rhoumenis* (Hesl.-Harr.f.) P.Delforge
DISTRIBUTION. Europe to C Siberia.
HABITAT. Grassland, meadows and marshes, sphagnum bogs, moorland, heathland, dune slacks, mostly on acidic soils; sea level to 2,400 m.

LEFT TO RIGHT
Denmark, Nordjylland 16.06.2004 (1)
Switzerland, Baselland 11.06.2008
England, Sussex 26.06.2022 (25)
Switzerland, Grisons 17.07.2002
Switzerland, Baselland 20.06.2003
Switzerland, Aargau 24.06.2001

FLOWERING. May to early August.

DISTINGUISHING FEATURES. Stem non-squeezable. Leaves green to grey-green, usually brown-spotted on upper surface; lowermost sheathing leaf acute. Bracts minutely crenate or serrate (observe the bract margin against the light using a 20× hand lens!). Flowers mauve to violet or white. Lateral sepals moderately to widely spreading. Lip shallowly 3-lobed, usually with well-defined markings (occasionally unmarked); mid-lobe distinctly smaller than side lobes; spur thinner than ovary.

NOTE. Although being clearly separated in most of their common geographic range, subsp. *maculata* and subsp. *fuchsii* can be virtually impossible to tell apart in certain regions – especially in C Europe and N Scandinavia.

LEFT TO RIGHT, TOP ROW
subsp. *maculata*
France, Aveyron 18.05.2003
Switzerland, Vaud 13.07.1991 (8)
France, Aveyron 18.05.2003

b. *Dactylorhiza maculata* subsp. *fuchsii* (Druce) Hyl., *Nord. Kärlväxtfl.* 2: 387 (1966).
Common spotted-orchid

SYNONYMS. *Orchis fuchsii* Druce; *Dactylorhiza fuchsii* (Druce) Soó; *D. fuchsii* subsp. *okellyi* (Druce) Soó; *D. fuchsii* subsp. *psychrophila* (Schltr.) Holub; *D. psychrophila* (Schltr.) Aver.; *D. andoeyana* Perko

DISTRIBUTION. Europe and temperate Asia across to Mongolia.

HABITAT. In full sun in tundra, peat bogs, grassland, marshes, meadows, dune slacks, open woods and woodland margins, usually on calcareous soils; sea level to 2,400 m.

FLOWERING. May to early August.

DISTINGUISHING FEATURES. Differs from the other three subspecies by the following combination of features: leaves usually brown-spotted on upper surface; lowermost sheathing leaf rounded to obtuse; lower bracts shorter than the flowers (very rarely longer); lip moderately to deeply (rarely shallowly) 3-lobed with the mid-lobe mostly about as large as the side lobes; spur thinner than the ovary.

NOTE. *Dactylorhiza maculata* subsp. *hebridensis* (Wilmott) H.Baumann & Künkele is a name that is often recognised for populations of generally low and condensed plants with dense inflorescences and dark-coloured flowers in the Hebrides and W Ireland. However, '*hebridensis*' does not seem to be clearly separated from subsp. *fuchsii*.

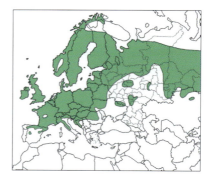

LEFT TO RIGHT, FROM MIDDLE ROW
subsp. *fuchsii*
Russia, Leningrad, Region 5 28.06.2009 (46)
Switzerland, Schwyz 17.06.1999 (2)
France, Cantal 28.06.2009
Switzerland, Obwalden 03.06.2006
France, Lozère 01.06.1989 (1)
Switzerland, Grisons 20.07.2001
Switzerland, Obwalden 06.06.1996
Germany, Black Forest 19.06.1998

150 | DACTYLORHIZA

c. *Dactylorhiza maculata* subsp. *maurusia* (Emb. & Maire) Soó, *Nom. Nov. Gen.* Dactylorhiza: 8 (1962).

SYNONYMS. *Orchis maurusia* Emb. & Maire; *Dactylorhiza battandieri* Raynaud; *D. maculata* subsp. *battandieri* (Raynaud) H.Baumann & Künkele

DISTRIBUTION. N Morocco and N Algeria.

HABITAT. Seepage fens, meadows and along brooks; 900–1,800 m.

FLOWERING. May to July.

DISTINGUISHING FEATURES. Differs from the typical subspecies in having consistently unspotted leaves and spurs that are thicker than the ovaries.

d. *Dactylorhiza maculata* subsp. *saccifera*
(Brongn.) Dilic, *Fl. SR Srbije* 8: 77 (1976).

SYNONYMS. *Orchis saccifera* Brongn.; *Dactylorhiza saccifera* (Brongn.) Soó; *D. saccifera* subsp. *bithynica* (H.Baumann) Kreutz; *D. saccifera* subsp. *gervasiana* (Tod.) Kreutz

DISTRIBUTION. Turkey, Romania and Mediterranean Europe west to Corsica and Sardinia.

HABITAT. Spring swamps, wet meadows, river beds and humid forests on alkaline rich soil; 100–2,200 m.

FLOWERING. May to July.

DISTINGUISHING FEATURES. Differs from the other three subspecies in having the following combination of features: leaves usually brown-spotted on upper surface; lower bracts longer than the flowers; lip moderately to deeply 3-lobed with mid-lobe approximately as large as the side lobes; spur thicker than the ovary.

NOTE. *Dactylorhiza amblyoloba* (Nevski) Aver. from the Caucasus and *D. phoenissa* (B.Baumann & H.Baumann) P.Delforge from Lebanon are similar to *D. maculata* subsp. *saccifera*, but have more shallowly lobed lips and generally less-spotted leaves. They might deserve taxonomic recognition at infraspecific level.

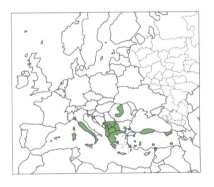

CLOCKWISE FROM TOP LEFT

subsp. *maurusia*
Morocco, Ketama 25.05.1975 (1)
Morocco, Ketama 25.05.1975 (1)

subsp. *saccifera*
Greece 11.07.1975 (9)
Greece 30.07.1977 (1)
Greece 11.07.1975 (9)

Dactylorhiza majalis (Rchb.) P.F.Hunt & Summerh., *Watsonia* 6: 130 (1965).
SYNONYM. *Orchis majalis* Rchb.

a. subsp. *majalis* Broad-leaved marsh-orchid
SYNONYM. *Dactylorhiza alpestris* (Pugsley) Aver.
DISTRIBUTION. Europe from the Ukraine to Denmark, France and NE Spain.
HABITAT. In full sun in damp to wet grassland, moors and alpine meadows; sea level to 2,600 m.
FLOWERING. May to July.
DISTINGUISHING FEATURES. Stem squeezable; upper two-thirds (inflorescence not included) bearing fewer nodes of sheathing leaves than lowermost third. Leaves green, suberect to somewhat spreading (angle between stem and longest leaf normally > 30°), often spotted on upper surface; longest leaf > 1.7 cm wide (except in dwarf individuals), apex flat. Bracts minutely crenate or serrate (observe the bract margin against the light using a 20× hand lens!). Flowers (purple-)mauve, rarely rose, white or purple. Lip 3-lobed with recurved to reflexed sides and acuminate to rounded (rarely truncate) mid-lobe; spur thicker than ovary, cylindrical or nearly so in lower view, < 11 mm long (except in giant individuals); lip of lowermost flower, when flattened, more than twice as wide as the stem just below the inflorescence (except sometimes in giant individuals).
NOTE. Broadly morphologically overlapping with the Irish *D. majalis* subsp. *occidentalis*. However, the two taxa have non-overlapping geographic ranges and are more genetically different than several morphologically well-separated subspecies of *D. majalis*.

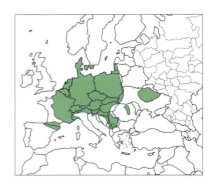

LEFT TO RIGHT, FROM TOP
subsp. *majalis*
Denmark, Sjælland 18.05.2023 (45)
Switzerland, Grisons 15.06.2013 (25)
France, Aveyron 17.07.2011
Switzerland, Schwyz 11.06.2008
Switzerland, Grisons 01.07.2017
France, Aveyron 12.06.2010
subsp. *baltica*
Russia, Leningrad 02.07.2008 (46)
Russia, Leningrad 12.06.2007 (46)
Russia, Leningrad 10.06.2008 (46)

b. *Dactylorhiza majalis* subsp. *baltica* (Klinge) H.Sund., *Europ. Medit. Orchid.* (Ed. 3): 40 (1980).
SYNONYM. *Orchis latifolia* subsp. *baltica* Klinge; *Dactylorhiza baltica* (Klinge) N.I.Orlova
DISTRIBUTION. From the Baltic region to Mongolia.
HABITAT. In full sun in wet meadows and moors; sea level to 400 m.
FLOWERING. June and July.
DISTINGUISHING FEATURES. Differs from the other 16 subspecies in having the following combination of features: stem squeezable; leaves green and suberect to somewhat spreading (angle between stem and longest leaf normally > 30°), with small to large, usually strong and evenly distributed brown spots on the upper surface; longest leaf > 1.7 cm wide (except in dwarf individuals) and with flat apex; flowers

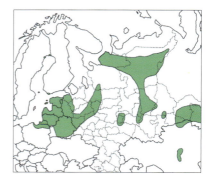

(purple-)mauve, rarely rose or white; lip with more or less spreading sides; spur cylindrical or nearly so in lower view, slightly to strongly downcurved and < 11 mm long (except in giant individuals).

NOTE. *Dactylorhiza ruthei* (M.Schulze ex Ruthe) Soó subsp. *ruthei*, reported from a few sites in the Baltic region, is genetically similar to *D. majalis* subsp. *baltica*, but differs in having consistently unspotted leaves and lips with much-reduced, often obscure markings. More comprehensive molecular data are needed to clarify the systematic position of *D. ruthei* subsp. *ruthei*.

c. *Dactylorhiza majalis* subsp. *calcifugiens*
H.A.Pedersen, *Nordic J. Bot.* 22: 655 (2004).
DISTRIBUTION. NW Denmark.
HABITAT. In full sun in old, moist to wet dune slacks with poor acid soil.
FLOWERING. June and July.
DISTINGUISHING FEATURES. Similar to subsp. *sphagnicola* but differs in having spreading leaves (angle between stem and longest leaf normally > 40°) and pure white, unmarked flowers with a spur that is usually < 9 mm long.

d. *Dactylorhiza majalis* subsp. *cordigera* (Fr.)
H.Sund., *Europ. Medit. Orchid.* (Ed. 2): 45 (1975).
SYNONYMS. *Orchis cordigera* Fr.; *Dactylorhiza cordigera* (Fr.) Soó; *D. cordigera* subsp. *bosniaca* (Beck) Soó
DISTRIBUTION. SE Europe (Carpathians and Balkans) to the Ukraine.
HABITAT. In full sun in damp grassland to wet marshes and meadows, on either acidic or alkaline soils; 900–2,400 m.
FLOWERING. May to early August.
DISTINGUISHING FEATURES. Differs from the other 16 subspecies in having the following combination of features: stem squeezable; leaves green and usually spotted on upper surface; flowers mauve to purple; lip shallowly 3-lobed with spreading to slightly recurved sides; spur distinctly conical in lower view.

NOTE. *Dactylorhiza graeca* H.Baumann from N Greece differs from typical *D. majalis* subsp. *cordigera* in having generally longer spurs and relatively narrower leaves. Molecular data (Hedrén et al., 2007) suggest that *D. graeca* is an aberrant population of *D. majalis* subsp. *cordigera*, possibly modified through limited and unidirectional gene flow from pollen of *D. majalis* subsp. *macedonica*.

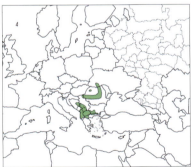

LEFT TO RIGHT, FROM TOP
subsp. *baltica*
Poland 25.06.1994 (1)
Poland 22.06.1999 (5)
subsp. *calcifugiens*
Denmark, Nordjylland 12.06.2002 (45)
Denmark, Nordjylland 11.06.2002 (45)
Denmark, Nordjylland 11.06.2002 (45)
subsp. *cordigera*
Greece 04.07.1977 (1)
Greece 18.07.1987 (1)
Romania 11.08.1980 (1)

e. *Dactylorhiza majalis* subsp. *elatior* (Fr.) Hedrén & H.A.Pedersen, *Nordic J. Bot.* 30: 271 (2012).

SYNONYMS. *Orchis latifolia* subsp. *elatior* Fr.; *Dactylorhiza osiliensis* Pikner; *D. ruthei* subsp. *osiliensis* (Pikner) Eccarius

DISTRIBUTION. The Baltic islands of Gotland, Saaremaa and Hiiumaa.

HABITAT. In full sun in moist meadows and marshes, on alkaline soils; sea level to 50 m.

FLOWERING. June and July.

DISTINGUISHING FEATURES. Differs from the other 16 subspecies in having the following combination of features: stem squeezable; leaves unspotted; flowers (purple-)mauve; lip with incurved or spreading (to recurved) sides and highly contrasting markings that extend far onto the side lobes; spur cylindrical or nearly so in lower view.

f. *Dactylorhiza majalis* subsp. *integrata* (E.G.Camus) H.A.Pedersen, *J. Eur. Orchid.* 41: 504 (2009). **Southern marsh-orchid**

SYNONYMS. *Orchis incarnata* var. *integrata* E.G.Camus; *Dactylorhiza praetermissa* (Druce) Soó; *D. praetermissa* var. *junialis* (Verm.) Senghas; *D. praetermissa* var. *integrata* (E.G.Camus) D.Tyteca & Gathoye

DISTRIBUTION. England, Wales and NW mainland Europe.

HABITAT. Wet meadows, dune valleys, on calcareous rich and wet soil; sea level to 600 m.

FLOWERING. May to July.

DISTINGUISHING FEATURES. Differs from the other 16 subspecies in having the following combination of features: usually > 25 cm tall; stem squeezable and > 3.5 mm in diameter just below inflorescence (except in dwarf individuals); sheathing leaves usually ≥ 4 in number, green, suberect (angle between stem and longest leaf usually 30–45°), unspotted or with small spots that merge to form annular patterns on upper surface; longest leaf > 1.7 cm wide (except in dwarf

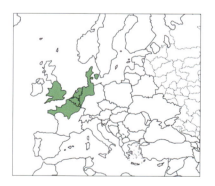

LEFT TO RIGHT, FROM TOP
subsp. *elatior*
Sweden, Gotland 24.06.2006 (45)
Sweden, Gotland 24.06.2006 (45)
Sweden, Gotland 25.06.2006 (45)
subsp. *integrata*
England, Devon 25.06.1994 (1)
England, Hampshire 08.07.2010 (25)
England, Hampshire 15.06.2004 (45) (var. *junialis*)

DACTYLORHIZA | 157

individuals) and with flat apex; inflorescence dense; flowers purple-mauve to rose or rarely white; lip with spreading sides; spur cylindrical in lower view, straight to slightly downcurved and < 11 mm long (except in giant individuals).

NOTE. *Dactylorhiza praetermissa* subsp. *schoenophila* R.M.Bateman & Denholm, described from bog-rush fens in E and S England, consists of populations with DNA profiles that accord with those of *D. majalis* subsp. *integrata*. However, it differs morphologically in being generally smaller with fewer and narrower, often apically hooded, leaves and in that the sides of the lip are moderately to strongly recurved. It might well deserve recognition at varietal level.

g. *Dactylorhiza majalis* subsp. *kalopissii*
(E.Nelson) H.A.Pedersen, P.J.Cribb & Rolf Kühn in Rolf Kühn *et al.*, *Field Guide Orchids Europe Medit.*: 154 (2019).

SYNONYM. *Dactylorhiza kalopissii* E.Nelson
DISTRIBUTION. NW Greece and S Albania.
HABITAT. In full sun in wet meadows, moors and marshes, on neutral to slightly acidic soils; 600–1,700 m.
FLOWERING. May to early July.
DISTINGUISHING FEATURES. Differs from the other 16 subspecies in having the following combination of features: stem squeezable; sheathing leaves usually widest below the middle and sometimes with tiny weak spots on the upper surface; flowers mauve or rarely white; lateral sepals finely dotted; lip shallowly to moderately 3-lobed, widest above the middle, gradually narrowed towards the base and usually < 15 mm wide; lip markings mainly consisting of distinct dots and never incorporating loops; spur cylindrical or nearly so in lower view.

TOP ROW, LEFT TO RIGHT
subsp. *integrata*
England, Hampshire 15.06.2004 (45)
Denmark, Fyn 28.06.2009 (45)
England, Hampshire ('*schoenophila*') 15.06.2004 (45)
Denmark, Nordjylland 14.07.2011 (45)

OPPOSITE, CLOCKWISE FROM LEFT
subsp. *kalopissii*
Greece 28.06.1977 (1)
Greece 28.06.1977 (1)
Greece 07.06.1999 (1)

h. *Dactylorhiza majalis* subsp. *lapponica*
(Laest. ex Hartm.) H.Sund., *Europ. Medit. Orchid.* (Ed. 2): 45 (1975).

SYNONYMS. *Orchis angustifolia* var. *lapponica* Laest. ex Hartm.; *O. latifolia* var. *dunensis* Rchb.f.; *Dactylorhiza lapponica* (Laest. ex Hartm.) Soó; *D. traunsteineri* (Saut. ex Rchb.) Soó; *D. traunsteineri* subsp. *curvifolia* (F.Nyl.) Soó; *D. curvifolia* (F.Nyl.) Czerep.; *D. schurii* (Klinge) Aver.; *D. traunsteineri* subsp. *carpatica* Batoušek & Kreutz; *D. carpatica* (Batoušek & Kreutz) P.Delforge; *D. lapponica* subsp. *angustata* (Arv.-Touv.) Kreutz; *D. traunsteineri* subsp. *turfosa* (F.Proch.) Kreutz; *D. traunsteineri* subsp. *bohemica*; *D. devillersiorum* P.Delforge; *D. traunsteineri* subsp. *irenica* (F.M.Vázquez) Kreutz.

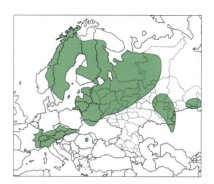

DISTRIBUTION. Disjunct areas in mainland Europe and across Siberia to Lake Baikal.

HABITAT. In full sun in wet montane meadows, marshes, alpine meadows and on stream banks, on alkaline soils; sea level to 2,150 m.

FLOWERING. Late May to July.

DISTINGUISHING FEATURES. Differs from the other 16 subspecies in having the following combination of features: stem usually squeezable; leaves more or less recurved, green to grey-green, spotted on the upper surface and normally with a flat apex; longest leaf < 1.7 cm wide (except in giant individuals); inflorescence lax (axis visible); flowers (purple-) mauve, rarely rose or white; lip with recurved (to slightly incurved) sides; spur cylindrical or nearly so in lower view, straight or nearly so.

NOTE. Broadly morphologically overlapping with *D. majalis* subsp. *traunsteinerioides* from the British Isles. However, the two taxa have non-overlapping geographic ranges and seem well separated genetically. British plants that are reported to belong to *D. majalis* subsp. *lapponica* in reality are subsp. *traunsteinerioides*.

LEFT TO RIGHT, FROM TOP
Switzerland, Obwalden 08.06.2002
Russia, Leningrad Region 3 21.06.2008 (46)
Sweden, Gotland 26.06.2006 (45)
Switzerland, Obwalden 08.06.2005
Switzerland, Obwalden 15.06.1983 (8)
Switzerland, Obwalden 11.06.2007
Switzerland, Obwalden 06.06.1996
Russia, Leningrad Region 1 14.06.2007 (46)
Switzerland, Obwalden 20.06.2009
Switzerland, Obwalden 29.06.1995
Russia, Leningrad Region 2 21.06.2008 (46)

i. *Dactylorhiza majalis* subsp. *macedonica*

(J.Hölz. & Künkele) H.A.Pedersen, P.J.Cribb & Rolf Kühn in Rolf Kühn *et al.*, *Field Guide Orchids Europe Medit.*: 158 (2019).

SYNONYM. *Dactylorhiza macedonica* J.Hölz. & Künkele; *Dactylorhiza kalopissii* subsp. *macedonica* (J.Hölz. & Künkele) Kreutz

DISTRIBUTION. From Bulgaria to SE Albania and N Greece.

HABITAT. Open wet grassland, on nutrient-rich soil; 1,000–1,500 m.

FLOWERING. May to early July.

DISTINGUISHING FEATURES. Similar to subsp. *kalopissii* but differs in having unmarked sepals and lip markings that are obscure or even absent. Besides, it also has a generally more dominant mid-lobe to the lip.

BELOW, LEFT TO RIGHT
Greece, Menikion 13.07.1987 (1)
Greece, Menikion 13.07.1987 (1)
Greece 07.06.1999 (2)

j. *Dactylorhiza majalis* subsp. *nieschalkiorum*

(H.Baumann & Künkele) H.A.Pedersen, P.J.Cribb & Rolf Kühn in Rolf Kühn *et al.*, *Field Guide Orchids Europe Medit.*: 159 (2019).

SYNONYM. *Dactylorhiza nieschalkiorum* H.Baumann & Künkele.

DISTRIBUTION. NW Anatolia.

HABITAT. In full sun and light shade in damp to wet marshes and flushes in open woodland in the coastal mountains, on calcareous and neutral soils; 1,000–1,800 m.

FLOWERING. June and July.

DISTINGUISHING FEATURES. Differs from the other 16 subspecies in having the following combination of features: stem usually squeezable; leaves green and unspotted (rarely spotted on upper surface); flowers mauve to purple; lip widest at or below the middle, not gradually narrowed towards the base, usually >15 mm wide; spur cylindrical or nearly so in lower view.

BELOW, LEFT TO RIGHT
Turkey, Bolu 01.07.1982 (1)
Turkey, Bolu 01.07.1982 (1)
Turkey, Abant Gölu 06.07.1977 (9)

k. *Dactylorhiza majalis* subsp. *occidentalis*
(Pugsley) P.D.Sell, *Acta Fac. Rerum Nat. Univ. Comen., Bot.* 14: 19 (1968).

SYNONYMS. *Orchis majalis* var. *occidentalis* Pugsley; *Dactylorhiza kerryensis* (Wilmott) P.F.Hunt & Summerh.; *D. kerryensis* var. *occidentalis* (Pugsley) Jebb.

DISTRIBUTION. Ireland.

HABITAT. In full sun in wet dune slacks, marshes, meadows and coastal fens; sea level to 200 m.

FLOWERING. Late May to July.

DISTINGUISHING FEATURES. Differs from the other 16 subspecies in having the following combination of features: usually < 25 cm tall; stem squeezable and > 3.5 mm in diameter just below the inflorescence (except in dwarf individuals); sheathing leaves usually ≥ 4 in number, normally spreading (angle between stem and longest leaf usually > 45°) and often spotted on the upper surface; longest leaf usually > 1.5 cm wide; inflorescence dense (axis hardly visible); flowers mauve to rose, rarely white or purple; lip with recurved to reflexed sides.

NOTE. Broadly morphologically overlapping with *D. majalis* subsp. *majalis* from mainland Europe. However, the two taxa have non-overlapping geographic ranges and differ genetically more than several morphologically well-separated subspecies of *D. majalis*.

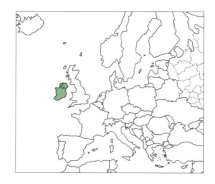

BOTH IMAGES
Eire, Connacht 22.05.1989 (22)

I. *Dactylorhiza majalis* subsp. *pindica* (B.Willing & E.Willing) H.A.Pedersen, P.J.Cribb & Rolf Kühn in Rolf Kühn *et al.*, *Field Guide Orchids Europe Medit.*: 161 (2019).

SYNONYMS. *Dactylorhiza pindica* B.Willing & E.Willing; *D. baumanniana* J.Hölz. & Künkele; *D. smolikana* B.Willing & E.Willing; *D. cordigera* subsp. *pindica* (B.Willing & E.Willing) H.Baumann & R.Lorenz; *D. baumanniana* subsp. *smolikana* (B.Willing & E.Willing) H.Baumann & R.Lorenz

DISTRIBUTION. N and W Greece south to N Peloponnese.

HABITAT. In full sun to light shade in damp or wet alkaline to slightly acidic soils in marshes and wet meadows and by streams; 1,000–2,000 m.

FLOWERING. Late May to early July

DISTINGUISHING FEATURES. Differs from the other 16 subspecies in having the following combination of features: stem squeezable; sheathing leaves usually widest below the middle and usually spotted on the upper surface; flowers (purple-)mauve; lip shallowly 3-lobed, widest above the middle, gradually narrowed towards the base and usually < 15 mm wide; lip markings dominated by lines and loops; spur cylindrical or nearly so in lower view.

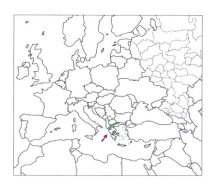

Greece, Epirus
14.06.1998 (45)

Greece, Epirus
14.06.1998 (45)

Greece, C Greece
09.06.1998 (45)

m. *Dactylorhiza majalis* subsp. *purpurella*

(T. & T.A.Stephenson) D.M.Moore & Soó, *Bot. J. Linn. Soc.* 76: 367 (1978). **Northern marsh-orchid**

SYNONYMS. *Orchis purpurella* T. & T.A.Stephenson; *Dactylorhiza purpurella* (T. & T.A.Stephenson) Soó

DISTRIBUTION. British Isles, Faroe Islands, W Norway, Denmark and The Netherlands.

HABITAT. In full sun in open moist grassland, clifftops, wet meadows, dune slacks, stream banks and swamps fed by springs on neutral to alkaline soil; sea level to 600 m.

FLOWERING. June to early August.

DISTINGUISHING FEATURES. Differs from the other 16 subspecies in having the following combination of features: stem squeezable; leaves green and sometimes spotted on the upper surface; flowers deep purple (rarely paler); lip with more or less incurved sides; spur cylindrical or nearly so in lower view; ovaries and inflorescence axis sometimes finely brown-dotted.

NOTE. The typical form has unspotted ovaries and inflorescence axis, and leaves with tiny or no spots. Plants having heavily spotted leaves, but unspotted ovaries and inflorescence axis, can be referred to as var. *maculosa* (T.Stephenson) R.M.Bateman & Denholm. Plants having spotted ovaries and inflorescence axis, and spotted or unspotted leaves, can be referred to as var. *cambrensis* (R.H.Roberts) R.M.Bateman & Denholm. If these varieties are recognised, the correct variety name for the typical form is var. *pulchella* (Druce) R.M.Bateman & Denholm.

LEFT TO RIGHT, FROM TOP

subsp. *purpurella* var. *pulchella*
Denmark, Nordjylland 02.07.1997 (45)
England 28.07.1986 (22)
England, County Durham 28.07.1985 (22)

BOTTOM ROW, FROM LEFT

subsp. *purpurella* var. *maculosa*
Wales, Ceredigion 18.06.1999 (45)

subsp. *purpurella* var. *cambrensis*
Denmark, Nordjylland 13.06.2007 (45)
Denmark, Nordjylland 11.06.2007 (45)
Denmark, Nordjylland 11.07.1998 (45)

168 | DACTYLORHIZA

n. *Dactylorhiza majalis* subsp. *pythagorae*
(Gölz & H.R.Reinhard) H.A.Pedersen, P.J.Cribb & Rolf Kühn in Rolf Kühn *et al.*, *Field Guide Orchids Europe Medit.*: 165 (2019).

SYNONYMS. *Dactylorhiza pythagorae* Gölz & H.R.Reinhard.

DISTRIBUTION. Mount Ambelos on Samos (Greece).

HABITAT. In full sun or light shade in seepages and by brooks in woodland, on alkaline soils; 600–900 m.

FLOWERING. Late May to early July.

DISTINGUISHING FEATURES. Differs from the other 16 subspecies in having the following combination of features: stem non-squeezable; sheathing leaves widest above the middle and unspotted; flowers pale mauve; lip shallowly to moderately 3-lobed, widest above the middle, gradually narrowed towards the base and usually < 15 mm wide; spur cylindrical or nearly so in lower view.

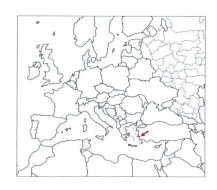

LEFT TO RIGHT, FROM TOP
subsp. *pythagorae*
Samos 07.07.1991 (1)
Samos 07.07.1991 (1)
subsp. *sesquipedalis*
France, Aveyron 20.05.1993
France, Aveyron 16.05.2007
France, Aveyron 25.05.1997
France, Aveyron 03.06.1989 (11)

o. *Dactylorhiza majalis* subsp. *sesquipedalis*
(Willd.) H.A.Pedersen & Hedrén, *Bot. J. Linn. Soc.* 168: 191 (2012).

SYNONYMS. *Orchis sesquipedalis* Willd.; *Dactylorhiza elata* (Poir.) Soó; *D. elata* var. *iberica* (T.Stephenson) Soó; *D. elata* subsp. *sesquipedalis* (Willd.) Soó; *D. elata* var. *durandii* (Boiss. & Reut.) Landwehr; *D. elata* var. *sesquipedalis* (Willd.) Landwehr; *D. elata* subsp. *ambigua* (Martrin-Donos) Kreutz; *D. atlantica* Kreutz & Vlaicha; *D. elata* var. *occitanica* (Geniez, Melki, Pain & Soca) P.Delforge; *D. elata* var. *algerica* (Rchb.f.) Delforge; *D. elata* var. *elongata* (Maire) Kreutz; *D. elata* subsp. *atlantica* (Kreutz & Vlaicha) Eccarius

DISTRIBUTION. SW Netherlands, W France, Iberian Peninsula and NW Africa, also in Corsica and Sicily.

HABITAT. In full sun in fens, wet meadows, coastal marshes and seepages, montane meadows, road banks and moors on alkaline substrates; sea level to 3,000 m.

FLOWERING. Late April to early August.

DISTINGUISHING FEATURES. Differs from the other 16 subspecies in having the following combination of features: stem squeezable; leaves green, erect (angle between stem and longest leaf normally < 30°) and unspotted or rarely with weak spots on the upper surface; longest leaf > 1.7 cm wide (except in dwarf individuals) and with flat apex; flowers (purple-) mauve, rarely rose or white; lip with spreading to recurved sides; spur cylindrical or nearly so in lower view and > 11 mm long (except in dwarf individuals).

NOTE. Often confused in cultivation with *D. foliosa* but differs in having a squeezable stem and a more cylindrical inflorescence of richer (purple-) mauve flowers with distinct dark purple lip markings. *Dactylorhiza majalis* subsp. *brennensis* (E.Nelson) D.Tyteca & Hedrén from C France has the appearance of a delicate, small-flowered *D. majalis* subsp. *elata*. Studying the variation in nuclear and plastid molecular markers, Hedrén & Tyteca (2020) demonstrated that the former taxon has originated through a genetic merger of populations of subsp. *elata* and subsp. *integrata*. Given the lack of clearly distinguishing traits against especially subsp. *elata*, we prefer to treat "*brennensis*" as a hybrid swarm that does not qualify for formal taxonomic recognition.

p. *Dactylorhiza majalis* subsp. *sphagnicola*
(Höppner) H.A.Pedersen & Hedrén, *Nordic J. Bot.* 22: 651 (2004).

SYNONYMS. *Orchis sphagnicola* Höppner; *Dactylorhiza sphagnicola* (Höppner) Soó

DISTRIBUTION. NE France to S Norway, Sweden and SE Finland.

HABITAT. In full sun on *Sphagnum*-rich moors, on acidic peaty soils; sea level to 600 m.

FLOWERING. Late May to early July.

DISTINGUISHING FEATURES. Differs from the other 16 subspecies in having the following combination of features: stem squeezable; leaves erect (angle between stem and longest leaf normally < 40°), straight to slightly incurved or recurved, green, unspotted (rarely with tiny, mostly weak spots on upper surface), apex boat-shaped; longest leaf < 1.7 cm wide (except in giant individuals); inflorescence dense; flowers mauve to rose with markings on lip (very rarely white and devoid of lip markings); lip with recurved or spreading (to slightly incurved) sides; spur cylindrical or nearly so in lower view, more or less downcurved and usually > 9 mm long.

OPPOSITE, LEFT TO RIGHT
subsp. *sphagnicola*
Belgium, Hohes Venn 21.06.1994 (3)
Belgium, Hohes Venn 21.06.1994 (3)
Sweden, Gästrikland 04.07.2005 (45)
Denmark, Sydjylland 14.06.2004 (45)

DACTYLORHIZA

q. *Dactylorhiza majalis* subsp. *traunsteinerioides*
(Pugsley) R.M.Bateman & Denholm, *Watsonia* 14: 372 (1983).

SYNONYMS. *Orchis majalis* subsp. *traunsteinerioides* Pugsley; *Dactylorhiza traunsteinerioides* (Pugsley) Landwehr ex R.M.Bateman & Denholm; *D. traunsteinerioides* var. *francis-drucei* (Wilmott) F.M.Vázquez; *D. ebudensis* (Wief. ex R.M.Bateman & Denholm) P.Delforge; *D. traunsteinerioides* subsp. *francis-drucei* (Wilmott) R.M.Bateman & Denholm

DISTRIBUTION. Great Britain, Ireland.

HABITAT. In full sun in damp to wet grassland, fens, marshes and wet dune slacks on alkaline (to neutral) soils; sea level to 200 m.

FLOWERING. May to early July.

DISTINGUISHING FEATURES. Differs from the other 16 subspecies in having the following combination of features: stem usually squeezable and < 3.5 mm in diameter just below the inflorescence (except in giant individuals); sheathing leaves usually < 4 in number and normally spotted on the upper surface; longest leaf usually < 1.5 cm wide; inflorescence lax (axis visible) and often somewhat secund; flowers mauve to rose, rarely white or purple; lip with spreading to recurved sides.

NOTE. Broadly morphologically overlapping with *D. majalis* subsp. *lapponica* from mainland Europe. However, the two taxa have non-overlapping geographic ranges and seem to be well separated genetically.

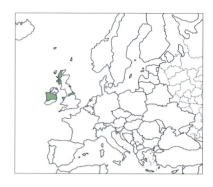

TOP ROW, FROM LEFT
subsp. *traunsteinerioides*
Wales, Anglesey 11.06.1999 (45)
Wales, Anglesey 11.06.1999 (45)

Dactylorhiza urvilleana (Steud.) H.Baumann & Künkele, *Mitt. Arbeitskreis Heimische Orchid. Baden-Württemberg* 13: 240 (1981).

SYNONYMS. *Orchis urvilleana* Steud.; *Dactylorhiza pontica* (Kohlmüller) P.Delforge; *D. urvilleana* subsp. *ilgazica* (Kreutz) Kreutz

DISTRIBUTION. N Anatolia to the Caucasus (with surroundings), Transcaucasia and N Iran.

HABITAT. In full sun to light shade in wet meadows, scrub and wet slopes, in alkaline soils; sea level to 2,700 m.

FLOWERING. May to August.

DISTINGUISHING FEATURES. 15–60(–90) cm tall. Stem non-squeezable. Leaves green to grey-green, usually brown-spotted on the upper surface. Bracts minutely

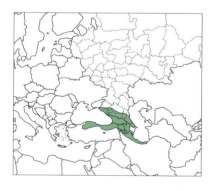

crenate or serrate (observe the bract margin against the light using a 20× hand lens!). Flowers mauve (to rose-coloured). Lip slightly 3-lobed, usually with well-defined markings; mid-lobe distinctly smaller than side lobes; spur > 3 mm in maximum diameter, thicker and usually longer than the ovary.

CLOCKWISE FROM TOP RIGHT
D. urvilleana
Turkey 20.05.2000 (45)
Turkey, Arvin 18.05.2000 (45)
Georgia 19.06.1996 (1)

25. CHAMORCHIS Rich., *De Orchid. Eur.*: 27 (1817).

A monotypic genus confined to Europe. Tiny plants that produce more than one tuber each year and often form colonies. The flowers are pollinated by ants, beetles and parasitic wasps.

Chamorchis alpina (L.) Rich., *Mém. Mus. Hist. Nat.* 4: 57 (1818).

SYNONYM. *Ophrys alpina* L.

DISTRIBUTION. The Scandinavian mountain chain, the Alps and the N & S Carpathians and at lower elevations in N Scandinavia and the Kola Peninsula.

HABITAT. In full sun in alpine meadows with *Dryas octopetala*, on dry calcareous soils; sea level (Norway) to 2,700 m (Alps).

FLOWERING. July–August.

DISTINGUISHING FEATURES. 5–10 cm tall. The leaves are grass-like and folded. The bracts are longer than the ovaries. The inflorescence has up to 15 small inconspicuous yellowish-green to brownish flowers. The spur-less, more or less entire, lip is 3–5 mm long. Sepals and petals form a helmet.

26. TRAUNSTEINERA Rchb., *Deut. Bot. Herb.-Buch:* 50 (1841).

A genus of two species, both found in our region. They have spherical tubers. Found in montane meadows, preferably, but not always, on calcareous soils and almost completely restricted to elevations above 1,000 m. Identification key on p. 428.

Traunsteinera globosa (L.) Rchb., *Fl. Sax.*: 87 (1844).
Globe orchid

SYNONYMS. *Orchis globosa* L., *Nigritella globosa* (L.) Rchb.

DISTRIBUTION. Mountains in C & S Europe from the Massif Central to the Crimea; also in the Caucasus.

HABITAT. Subalpine and alpine meadows on limestone; 500 to 2,700 m.

FLOWERING. June to August.

DISTINGUISHING FEATURES. 20–70 cm tall. It has lanceolate, suberect, blue-green leaves, a dense, globular or ovoid inflorescence with many small flowers that are usually pink with a purple-spotted lip, very rarely white. The sepals and petals are characteristically prolonged and swollen at the tips, the lateral sepals are usually less than 9 mm long. The bracts are as long as the ovary. The spur is shorter than the ovary.

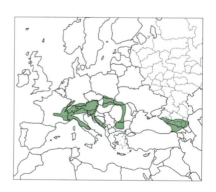

CLOCKWISE FROM LEFT
Switzerland, Bern 30.06.1986 (11)
France, Drôme 14.07.1990 (11)
Switzerland, Schwyz 20.06.2009
Switzerland, Baselland 20.06.2004
Italy, South Tyrol 12.07.2009

TRAUNSTEINERA 175

Switzerland, Grisons 15.07.1995   Switzerland, Grisons 19.07.2002   Switzerland, Grisons 17.07.2001
(inset) Switzerland, Grisons 18.07.2003

Turkey, Giresun 18.07.1982 (1)   Georgia, Borjomi 11.07.2014 (27)   Georgia, Borjomi 11.07.2014 (27)

Traunsteinera sphaerica (M.Bieb.) Schltr., *Repert. Spec. Nov. Regni Veg. Sonderbeih.* A 1: 227 (1928).
SYNONYMS. *Orchis sphaerica* M.Bieb.; *Traunsteinera globosa* subsp. *sphaerica* (M.Bieb.) Soó
DISTRIBUTION. The Crimea, the Caucasus and NE Turkey.
HABITAT. Montane meadows and open forests, in upper montane and subalpine area; 1,600–2,800 m.
FLOWERING. June to August.
DISTINGUISHING FEATURES. 20–70 cm tall. Differs from *T. globosa* in having consistently creamy white flowers that are also generally larger (lateral sepals usually more than 9 mm long).

27. ORCHIS L., *Sp. Pl.* 2: 939 (1753).

SYNONYM. *Aceras* R.Br.

A genus of 21 species characterised by ovoid tubers, a basal rosette of green or purple-spotted leaves and an erect spike of few to many usually showy flowers, the lip of which sometimes has scattered tufts of tiny, coloured hairs. The bracts are membranous and appressed to the ovaries. Most species are found in Europe, the Mediterranean region and the Middle East, but the generic range extends from E Europe across most of temperate Asia. Identification key on p. 428.

Orchis anthropophora (L.) All., *Fl. Pedem.* 2: 148 (1785). **Man orchid**
SYNONYMS. *Ophrys anthropophora* L.; *Serapias anthropophora* (L.) Jundz.
DISTRIBUTION. Widespread in W, C and Mediterranean Europe and the Mediterranean islands but absent from Ireland, Scotland, Wales, the Nordic countries and E Europe to the east of Germany; also in NW Africa, Cyprus and SW and S Turkey.
HABITAT. In open scrub, grassland and woodland and forest margins, on calcareous soils; sea level to 1,600 m.
FLOWERING. April to June.
DISTINGUISHING FEATURES. 10–40 cm tall. Easily recognised by its long slender spikes of greenish yellow to orange-brown flowers with a slender, deeply 3-lobed, spur-less lip.
NOTE. Previously separated in the genus *Aceras*, this species nests within the *Orchis* clade and hybridises readily with other *Orchis* species.

LEFT TO RIGHT
France, Aveyron, 20.05.2003
France, Aveyron, 22.05.2003
Crete 07.04.2014
Switzerland, Asp 07.06.2005
France, Aveyron 24.05.1996
France, Aveyron 18.05.2011

Orchis italica Poir. in Lamarck, *Encycl.* 4 (2): 600 (1798). **Naked-man orchid**

SYNONYM. *Orchis longicruris* Link.

DISTRIBUTION. Mediterranean Europe and NW Africa, across to Turkey and the Near East.

HABITAT. Seasonally wet meadows, garigue and open forest, on stony calcareous soil; sea level to 1,300 m.

FLOWERING. March–May.

DISTINGUISHING FEATURES. 20–50 cm tall. The leaves are characteristically undulate on the margins and sometimes dark spotted. The pink flowers, densely packed in a more or less pyramidal spike, have a man-like lip with slender tapering segments.

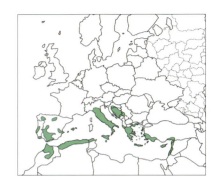

LEFT TO RIGHT FROM TOP
Crete 07.04.2014
Crete 05.04.2014
Crete 07.04.2014
Crete 05.04.2014
Crete 05.04.2014
Portugal, Coimbra 08.04.1992
Cyprus 16.03.2011

Orchis galilaea (Bornm. & M. Schulze) Schltr., *Repert. Spec. Nov. Regni Veg.* 19: 47 (1923). **Galilean orchid**

SYNONYMS. *Orchis punctulata* var. *galilaea* Bornm. & M. Schulze; *Orchis punctulata* subsp. *galilaea* (Bornm. & M.Schulze) Soó

DISTRIBUTION. Syria to Jordan, possibly only extant in Israel and Lebanon.

HABITAT. Oak and pine forest, and garigue; 300–1,000 m.

FLOWERING. February to April.

DISTINGUISHING FEATURES. 20–60 cm tall. Allied to *O. simia* with the flowers opening from the top of the inflorescence downwards but readily distinguished by the smaller pale yellow-green flowers with purple stripes on the sepal hood and dark purple spots on the base of the lip.

LEFT TO RIGHT
Israel 03.1989 (10)
Israel 14.03.1994 (1)
Israel 09.04.1981 (5)
Israel 04.1979 (21)

Orchis simia Lam., *Fl. Franç. (Lamarck)* 3: 507 (1779).
Monkey orchid

SYNONYMS. *Orchis militaris* subsp. *simia* (Lam.) Bonnier & Layens; *O. taubertiana* B. & H.Baumann; *O. simia* subsp. *taubertiana* (B. & H.Baumann) Kreutz

DISTRIBUTION. Mediterranean and W Europe, very rare in SE England, absent from Portugal and much of N and C Europe north of the alps; also in Turkey, the Crimea, and the Caucasus across to Iran.

HABITAT. Open forest, woodland and bush, nutrient-poor meadows, on moderate dry and calcareous soil; sea level to 1,800 m.

FLOWERING. April–June.

DISTINGUISHING FEATURES. 25–50 cm tall. Leaves green and shiny. The spike is densely flowered and flowers from the top down. The flowers are man-like with long slender but not tapering purple limbs.

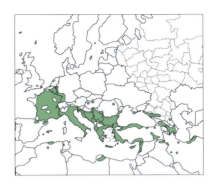

CLOCKWISE FROM TOP LEFT
France, Aveyron 18.05.2011
France, Aveyron 18.05.2011
Crete 12.04.2015
Italy, Tuscany 05.05.2007
France, Aveyron 18.05.2003

Orchis punctulata Steven ex Lindl., *Gen. Sp. Orchid. Pl.*: 273 (1835). **Punctate orchid**

SYNONYMS. *Orchis punctulata* var. *sepulchralis* Rchb.f.; *O. punctulata* subsp. *schelkownikowii* (Woronow) Soó

DISTRIBUTION. From easternmost Greece across Turkey to Cyprus, the Levant, the Caucasus, the Crimea and N Iran.

HABITAT. Semi-shade in meadows, garigue, on the margins of oak and pine woods and on river banks, on base-rich soil; sea level to 1,800 m.

FLOWERING. February to May.

DISTINGUISHING FEATURES. 25–70 cm tall. A distinctive orchid because of its yellow flowers with the lip often edged with red-brown on the lobes. The flowers are similar in shape to those of the closely related *O. militaris* and *O. purpurea*. The golden-yellow lip with strongly falcate side lobes, which are less than 3 times as long as wide, separates *O. punctulata* from *O. adenocheila*.

CLOCKWISE FROM LEFT
Cyprus, Neo Chorio 11.03.2006
Cyprus 14.02.1995 (14)
Cyprus, Nicosia 14.03.1999
Cyprus, Neo Chorio 11.03.2006
Cyprus, Neo Chorio 11.03.2006

Orchis adenocheila Czerniak., *Bot. Mat. Gerb. Sada RSFSR* 5: 173 (1924).

SYNONYMS. *Orchis militaris* var. *adenocheila* (Czerniak.) Hautz.; *Orchis punctulata* subsp. *adenocheila* (Czern.) Hautz.

DISTRIBUTION. Azerbaijan to N Iran.

HABITAT. A rare species of calcareous meadows, open woodland and scrub; 150–1,600 m.

FLOWERING. Early April to mid-June.

DISTINGUISHING FEATURES. Plant 20–60 cm tall. Similar to *O. punctulata* but differs in its creamy-white to pale rose or sulphur-yellow lip with straight to weakly falcate side lobes that are more than 3 times as long as broad. Besides, the sepals are green rather than weakly greenish yellow.

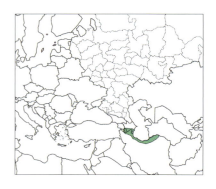

ALL IMAGES
Azerbaijan 03.05.1997 (57)

Orchis purpurea Huds., *Fl. Angl.*: 334 (1762).
Lady orchid

SYNONYM. *Orchis militaris* var. *purpurea* (Huds.) Huds.

a. subsp. *purpurea*

DISTRIBUTION. Widespread in Europe north to SE England and Denmark; also in Turkey, the Crimea and NE Algeria.

HABITAT. Open bush, garigue and woodland, forest margins and meadows, on moderately dry to damp calcareous soil; sea level to 2,000 m.

FLOWERING. April to June.

DISTINGUISHING FEATURES. 30–80 cm tall. It has green glossy leaves at the base; inflorescence stout and densely many-flowered in a cylindrical or ellipsoid spike; flowers with a green to dark purple-brown hood, formed by the sepals and petals, and a large

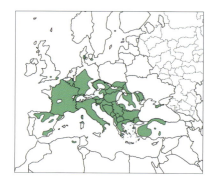

ABOVE, LEFT TO RIGHT
O. purpurea subsp. *purpurea*
France, Aveyron 23.05.2001
France, Aveyron 23.05.2005
France, Aveyron 23.05.2008
France, Aveyron 13.05.2008
France, Aveyron 21.05.2003
France, Aveyron 23.05.2008
France, Aveyron 23.05.2008
France, Aveyron 23.0.2003
subsp. *caucasica*
Turkey, Artvin 17.05.2000 (45)
Azerbaijan, Kuba 18.04.2000 (1)

tri-lobed lip with a fan-shaped mid-lobe, with purple-haired spots, and narrow spreading side lobes.

b. *Orchis purpurea* subsp. *caucasica* (Regel) B. & H.Baumann, R.Lorenz & R.Peter, *J. Eur. Orch.* 35 (1): 182 (2003).

SYNONYMS. *Orchis caucasica* Regel; *Orchis purpurea* var. *caucasica* (Regel) E.G.Camus

DISTRIBUTION. NE Turkey to the Caucasus and adjoining regions.

HABITAT. Deciduous and oak forest, forest margins, garigue, on moderate dry soil rich in basic elements; sea level to 1,500 m.

FLOWERING. May and June.

DISTINGUISHING FEATURES. 30–70 cm tall. Differs from the typical subspecies in having deep magenta (not green to dark purple-brown) sepals. Also, the flowers are generally smaller.

Orchis militaris L., *Sp. Pl.* 2: 941 (1753).
Military orchid

a. subsp. *militaris*

DISTRIBUTION. Widespread in Europe but only in mountains in the Mediterranean and absent from Iceland, most of Scandinavia, Ireland and Portugal; in Britain, confined to a few localities in S England; also in the northern Caucasus and across Siberia to Mongolia, W China and Afghanistan.

HABITAT. Seasonally wet sites in open forest and bush, moderately dry grassland and meadows, quarries and roadsides, on calcareous soil; sea level to 2,000 m.

FLOWERING. April–June.

DISTINGUISHING FEATURES. 20–60 cm tall. Leaves with flat margins, unspotted. The inflorescence flowers from the base upwards. The lip is characteristically man-shaped with relatively short broad limbs, a characteristic short pointed tail between the legs and tufts of short purple hairs along the middle. Angle between the lip and the sepal helmet in side view > 50°.

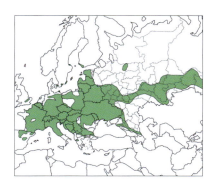

b. *Orchis militaris* subsp. *stevenii* (Rchb.f.) B.Baumann, R.Lorenz & Ruedi Peter, *J. Eur. Orch.* 35: 179 (2003).

SYNONYM. *Orchis stevenii* Rchb.f.

DISTRIBUTION. From the Crimea and Turkey to the southern Caucasus and N Iran.

HABITAT. In sun and light shade on grassland and in dry meadows, scrubs and open woodland on calcareous soil; 0–2,000 m.

FLOWERING. May to June.

DISTINGUISHING FEATURES. 20–50 cm, or more, tall. Differs from the typical subspecies in the angle between lip and sepal helmet in side view being < 50°. Also, it tends to have a laxer inflorescence and larger flowers with a narrower midlobe to the lip.

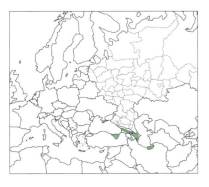

OPPOSITE, LEFT TO RIGHT
Orchis militaris subsp. *militaris*
France, Aveyron 20.05.2002
France, Alsace 09.05.2021
France, Burgundy, 25.05.2019 (25)
France, Alsace 09.05.2021
France, Burgundy, 27.05.2019 (25)
France, Alsace 09.05.2021
France, Dolmen 20.05.2000
France, Aveyron 21.05.2001
Switzerland, Aargau 22.05.1979 (11)
Orchis militaris subsp. *stevenii*
Azerbaijan, Qabala 10.05.2022 (56)
Azerbaijan, Qabala 10.05.2022 (56)
Ukraine, Crimea 24-05-2011 (56)

ORCHIS

Orchis spitzelii Saut. ex W.D.J. Koch, *Syn. Fl. Germ. Helv.* 1 (2): 686 (1837). **Spitzel's orchid**

SYNONYMS. *Orchis patens* subsp. *spitzelii* (Saut. ex W.D.J. Koch) Á.Löve & Kjellq.; *Barlia spitzelii* (Saut. ex W.D.J. Koch) Szlach.

a. subsp. *spitzelii*

SYNONYM. *Orchis latiflora* (B. & H.Baumann) P.Delforge

DISTRIBUTION. Scattered localities in C and S mainland Europe (west to the Pyrenees) and Corsica, and with a northern outpost in Gotland; Asia from Anatolia to Iran; also in N Algeria. A rare species that usually occurs in small numbers in any locality.

HABITAT. In glades in conifer and oak forest, mountain ledges, screes and meadows, on limestone and dolomite; sea level to 2,100 m.

FLOWERING. April to June.

DISTINGUISHING FEATURES. 15–35 cm, or more, tall. Leaves green, shiny. Inflorescence narrowly cylindrical. Flowers with short, more or less connivent, green to brownish-purple sepals and petals, the sepals with dark purple spots on the ventral side. Lip large, pink, heavily purple-spotted, deflexed, trilobed, with reflexed sides, an oblong, emarginate mid-lobe and a stout, conical, downcurved, obtuse spur, 6–10 mm long.

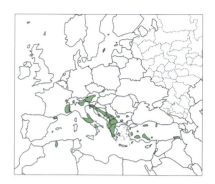

LEFT TO RIGHT FROM TOP
Switzerland, Valais 29.06.2001
Switzerland, Valais 29.06.2001
Turkey, Antalya 19.05.2013 (40)
Turkey, Antalya 19.05.2013 (40)
Turkey, Antalya 19.05.2013 (40)
Turkey, Antalya 19.05.2013 (40)

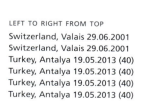

ORCHIS | 187

b. *Orchis spitzelii* subsp. *nitidifolia* (W.P.Teschner) Soó, *Bot. J. Linn. Soc.* 76 (4): 368 (1978).

SYNONYMS. *Orchis patens* subsp. *nitidifolia* W.P. Teschner; *O. prisca* Hautz.

DISTRIBUTION. Endemic to Crete.

HABITAT. In sun and light shade in glades in pine and maple forest and phrygana on dry calcareous soil; 600–1,600 m.

FLOWERING. April to June.

DISTINGUISHING FEATURES. 20–40 cm tall. Similar to the typical subspecies but differs in having pink sepals and petals, the lateral sepals widely spreading, with a green central stripe and often a few dark purple spots. Besides, the lip is more deeply lobed and has a well-defined white median band that is spotted with purple.

c. *Orchis spitzelii* subsp. *cazorlensis* (Lacaita) D. Rivera & López Vélez, *Orquid. Prov. Albacete*: 129 (1987).

SYNONYMS. *Orchis cazorlensis* Lacaita; *Androrchis spitzelii* subsp. *cazorlensis* (Lacaita) D.Tyteca & E.Klein

DISTRIBUTION. Mainland Spain, Majorca and N Morocco.

HABITAT. Clear coniferous forests, on soil rich in basic elements; sea level–1,900 m.

FLOWERING. May to June.

DISTINGUISHING FEATURES. 20–35 cm tall. Similar to the typical subspecies but differs in having brighter flowers with unspotted or few-spotted sepals and a lip with more or less spreading sides.

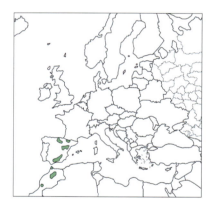

OPPOSITE, LEFT TO RIGHT
subsp. *nitidifolia*
Crete 20.05.1980 (12)
Crete 09.05.1997 (14)
Crete 09.05.1997 (14)
subsp. *cazorlensis*
Spain, Cuenca 26.05.2002 (15)
Spain, Cuenca 28.05.2002 (15)
Spain, Cuenca 24.05.2002 (15)
Spain, Cuenca 24.05.2002 (15)

Italy, Liguria 17.05.1996 (14) Italy, Liguria 17.05.1996 (14) Italy, Liguria 05.1980 (12)

Canary Islands, Gran Canaria
05.05.1995 (1)

Canary Islands, Tenerife
12.03.2023 (25)

Canary Islands, La Gomera
24.01.1995 (1)

Orchis patens Desf., *Fl. Atlant.* 2: 318 (1799).
Atlas orchid

a. subsp. *patens*

SYNONYMS. *Orchis brevicornis* Viv.; *O. kelleri* E.Hunz. ex G.Keller & Soo; *O. panormitana* Tineo.

DISTRIBUTION. Disjunct, comprising Liguria in Italy and N Algeria to N Tunisia.

HABITAT. Open areas in chestnut, oak and cedar forest, sparse grassland on sandstone; up to 700 m in Europe but 1,600 m in N Africa.

FLOWERING. March to June.

DISTINGUISHING FEATURES. 25–50 cm, or more, tall. Similar to *O. spitzelli* subsp. *nitidifolia* but differs in having an erect (not incurved) dorsal sepal and a more deeply trilobed lip with usually dentate lobes and a straight, apically rounded spur.

b. *Orchis patens* subsp. *canariensis* (Lindl.) Asch. & Graebn., *Orquid. Prov. Albacete*: 129 (1987).

SYNONYMS. *Orchis canariensis* Lindl.; *Barlia canariensis* (Lindl.) Szlach.

DISTRIBUTION. Canary Islands only.

HABITAT. Misty pine forest, tree-heather scrub, steep rocky canyons and ledges, on slightly acidic volcanic rocks; sea level to 1,800 m.

FLOWERING. March to June.

DISTINGUISHING FEATURES. 25–40 cm tall. A generally smaller plant than the typical subspecies with broader leaves, a dorsal sepal that bends over the column, lateral sepals that are less strongly marked with green, and a lip with a larger basal area of white and fewer spots.

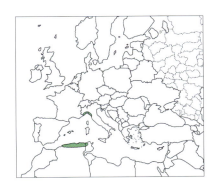

OPPOSITE
1st row: subsp. *patens*
2nd row: subsp. *canariensis*

Orchis sitiaca (Renz) P. Delforge, *Naturalistes Belges* 71 (3): 107 (1990).

SYNONYMS. *Orchis anatolica* subsp. *sitiaca* Renz; *Androrchis sitiaca* (Renz) D.Tyteca & E.Klein

DISTRIBUTION. Endemic to Cretan mountains.

HABITAT. Meadow-like areas in scrub and bush and in dry grassland, rarely in pine forest, on slightly acidic or neutral ground; up to 1,400 m.

FLOWERING. Late March to May.

DISTINGUISHING FEATURES. 10–40 cm tall. Similar to *O. anatolica* but differs in having silvery grey-green leaves and a deeply 3-lobed lip with narrower, reflexed side lobes.

LEFT TO RIGHT
Greece, Crete 05.04.2013 (25)
Crete 07.04.2014
Crete, Spili 16.04.2016

Orchis anatolica Boiss., *Diagn. Pl. Orient. ser.* 1 (5): 56 (1844). **Anatolian orchid**

SYNONYMS. *Orchis quadripunctata* subsp. *anatolica* (Boiss.) Asch. & Graebn; *O. anatolica* subsp. *troodi* (Renz) Renz; *O. troodi* (Renz) P.Delforge

DISTRIBUTION. From eastern Greece across Turkey and Cyprus to Israel and Iran.

HABITAT. Shade and partial shade in open deciduous and pine woods and phrygana, on dry and calcareous to weakly acidic soil; up to 2,400 m.

FLOWERING. March to May.

DISTINGUISHING FEATURES. 10–50 cm tall. Readily distinguished by its flowers which have an ascending or upcurved, 15–25 mm long, tapering spur. Its closest ally appears to be *O. sitiaca* from which it differs in having green, rather than silvery green, leaves and a shallowly (to moderately) 3-lobed lip, which is weakly folded along the mid-line.

NOTE. Plants from the Troodos Mountains in Cyprus have been distinguished by some authors as *O. troodi* but they fall within the range of the species and do not merit taxonomic recognition.

OPPOSITE PAGE, LEFT TO RIGHT
Cyprus, Akamas 18.03.2002
Cyprus, Akamas 16.03.2006
Cyprus, Akamas 24.03.1996
Cyprus, Troodos 16.03.2011
Cyprus, Akamas 24.03.2005
Cyprus, Akamas 24.03.2005
Cyprus, Troodos 18.03.1996
Cyprus, Troodos 18.03.2008
Cyprus, Akamas 18.03.2008

THIS PAGE, LEFT TO RIGHT
Cyprus, Troodos 15.03.2005
Cyprus, Troodos 17.03.2001
Cyprus, Troodos 18.03.2008
Crete, Thripti 20.05.1989 (14)
Cyprus, Kapilo 22.05.1990 (14)
Greece, Dodecanese 06.04.1966 (10)

LEFT TO RIGHT FROM TOP
O. quadripunctata
Crete, Kandanos 06.04.2014
Crete, Cato Saktuoria 05.04.2014
Crete, Kissos 07.04.2014
Crete, Kissos 07.04.2014
Crete, Cato Saktuoria 05.04.2014
Crete, Kissos 07.04.2014
Crete, Kissos 07.04.2014
Crete, Kissos 07.04.2014
Crete, Kissos 10.04.2014

Orchis quadripunctata Cirillo ex Ten., *Prod. Fl. Neap.* 1: 53 (1812). **Four-spotted orchid**

SYNONYM. *Androrchis quadripunctata* (Cirillo ex Ten.) D.Tyteca & E.Klein

DISTRIBUTION. S mainland Italy, former Yugoslavia, Albania, Greece with Crete and the Aegean Islands, Cyprus; very local in W Turkey.

HABITAT. Sunny spots in bush, open woodland, sparse grassland and on rocky slopes on limestone; sea level to 1,600 m.

FLOWERING. March to May.

DISTINGUISHING FEATURES. 10–30 cm tall. Leaves usually spotted. Sepals spreading, 4–7 mm long. Lip broader than long, larger than the sepals, 3-lobed with side lobes at least as large as the mid-lobe, and usually 2–5 spots at the mouth of the thin decurved spur.

NOTE. *Orchis* × *sezikiana* B.Baumann & H. Baumann from Cyprus probably represents old hybrid swarms between *O. anatolica* and *O. quadripunctata*.

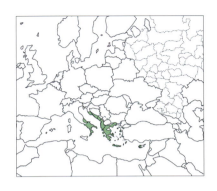

BELOW, LEFT TO RIGHT
O. ×*sezikiana*
Cyprus, Troodos 15.03.2009
Cyprus, Troodos 18.03.2011
Cyprus, Troodos 17.03.2008
Cyprus, Troodos 15.03.2006
Cyprus, Troodos 15.03.2006
Cyprus, Akamas 18.03.2011

Orchis brancifortii Biv., *Stirp. Rar. Sicil. i.* [11] t. 1. f. 1 (1813). **Brancifort's orchid**

SYNONYM. *Anacamptis brancifortii* (Biv.) Lindl.

DISTRIBUTION. Sardinia, Sicily and Calabria.

HABITAT. On rocky slopes and stony garigue, and in rock crevices, sparse grassland and open bush, on limestone; 200 to 1,700 m.

FLOWERING. April and June.

DISTINGUISHING FEATURES. 10–25 cm tall. Allied to *O. quadripunctata* and with similarly sized flowers, but differs in having consistently unspotted leaves and a remarkably small lip (never larger than the sepals) with a mid-lobe that is larger than the slender side lobes.

Orchis pallens L., *Mant. Pl. Altera*: 292 (1771). **Pale-flowered orchid**

SYNONYM. *Androrchis pallens* (L.) D.Tyteca & E.Klein

DISTRIBUTION. N Spain and S France, C Europe across to S Ukraine and south to Italy and Greece, absent from Ireland, Britain, the Nordic countries and N and E Europe; also in Turkey and the Caucasus.

HABITAT. Deciduous woodland, open pine forest margins, relatively dry meadows and mountain grassland, on limestone and rarely on sandstone; 100–2,400 m.

FLOWERING. April to early June.

DISTINGUISHING FEATURES. 20–40 cm tall. It differs from other yellow-flowered *Orchis* by the combination of broad, glossy green, unspotted leaves, densely arranged large flowers that smell of cats, and an unspotted lip with a stout spur dilating to the tip.

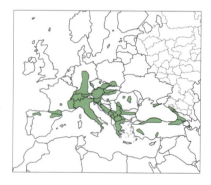

ORCHIS | 197

Sardinia 10.04.1982 (15)

Sicily 15.04.2006 (3)

Sicily 04.1982 (21)

Germany, Blumegg
28.04.1995

Germany, Blumegg
28.04.1995

Germany, Blumegg
6.04.1995

LEFT Germany, Blumegg 20.04.1999

Orchis provincialis Balb. ex Lam. & DC., *Syn. Fl. Gall.*: 169 (1806). **Provençal orchid**

SYNONYMS. *Orchis morio* var. *provincialis* (Balb. ex Lam. & DC.) Pollini; *Androrchis provincialis* (Balb. ex Lam. & DC.) D.Tyteca & E.Klein

DISTRIBUTION. Mediterranean Europe from Portugal, Spain, S France, S Switzerland, Italy and the larger islands of the Mediterranean across to the Balkans, Greece, the Crimea and N and W Turkey.

HABITAT. Open sunny deciduous and coniferous forests and thorn-scrub, on moderately acidic to neutral soil over sandstone, dolomite and serpentine; sea level to 1,700 m.

FLOWERING. April and May.

DISTINGUISHING FEATURES. 20–35 cm tall. Differs from other yellow-flowered *Orchis* species in having dark purplish black spotted leaves, pale yellow flowers and a lip that is convex from the side-view.

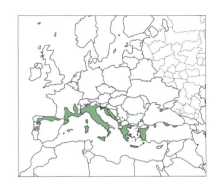

LEFT TO RIGHT
Sardinia, Fonni 12.04.2017
Sardinia, Fonni 12.04.2017
France, Liquisses 20.04.1998
Italy, Quirico d'Orcia 17.04.1996
France, La Couvertoirade 22.05.1996
Italy, Quirico d'Orcia 17.04.1996

Orchis laeta Steinh., *Ann. Sci. Nat., Bot.* (sér. 2) 9: 209 (1838).

DISTRIBUTION. NE Algeria and NW Tunisia.

HABITAT. Full sun to partial shade in open oak and cedar forests, mainly on siliceous soil; 300–1,800 m.

FLOWERING. March to early May.

DISTINGUISHING FEATURES. 15–25 cm tall. Leaves unspotted. Flowers yellowish white to pale yellow, or rose to pink. Lateral sepals reflexed to recurved and erect, devoid of markings. Lip moderately 3-lobed with purple dots on the basal to central part (sometimes extending onto the mid-lobe); side lobes recurved; spur slender, upcurved, 17–30 mm long, at least twice as long as the lip blade and distinctly longer than the ovary.

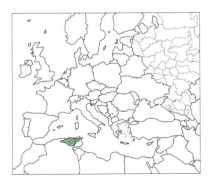

BOTH IMAGES
Algeria, Constantine 01.04.2004 (1)

Orchis olbiensis Reut. ex Gren., *Mém. Soc. Émul. Doubs* (sér. 3) 4: 404 (1860).

SYNONYM. *Orchis mascula* var. *olbiensis* (Reut. ex Gren.) Nyman

DISTRIBUTION. W Mediterranean from Portugal to S France and Corsica; also in NW Africa.

HABITAT. In sun on rocky slopes, sparse grassland, maquis and garigue, on limestone; up to 2,000 m.

FLOWERING. March to May.

DISTINGUISHING FEATURES. 10–30 cm tall. Similar to *O. mascula* subsp. *mascula* but differs in having a lip whose sides are strongly recurved and with spots that extend from the basal and central part to the mid-lobe and sometimes also to the side lobes.

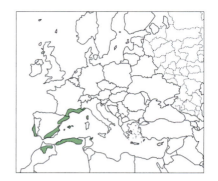

Orchis pauciflora Ten., *Fl. Napol. 1 (Prodr.): LII* (1811).
Sparse-flowered orchid

SYNONYMS. *Orchis provincialis* subsp. *pauciflora* (Ten.) Lindl.; *Androrchis pauciflora* (Ten.) D.Tyteca & E.Klein

DISTRIBUTION. Corsica, mainland Italy, Albania, former Yugoslavia, Ionian Islands, mainland Greece, Cyclades and Crete.

HABITAT. Open forest, garigue, scrub, sparse meadows, on calcareous soil; sea level to 1,800 m.

FLOWERING. March to May.

DISTINGUISHING FEATURES. 10–30 cm tall. Can be distinguished from the other yellow-flowered species by its unspotted leaves, few-flowered inflorescences of large flowers with a bright yellow lip (contrasting the pale sepals and petals), and the lip marked along its centre by two more or less distinct rows of red-brown dots.

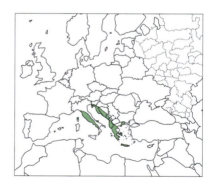

ORCHIS | 201

Spain, Malaga 21.03.1999 (14)

Spain, Malaga 21.03.1999 (14)

Mallorca 29.03.1989 (11)

Crete, Melambes 02.04.2014

Crete, Melambes 02.04.2014

Crete, Melambes 02.04.2014

Orchis mascula (L.) L., *Fl. Suec.* (Ed. 2): 310 (1755).
Early-purple orchid

SYNONYM. *Orchis morio* var. *mascula* L.

a. subsp. *mascula*

SYNONYMS. *Orchis pinetorum* Boiss. & Kotschy; *O. tenera* (Landwehr) Kreutz

DISTRIBUTION. Widespread in Europe across to Russia, also in NW Africa, Turkey and the Middle East.

HABITAT. In open deciduous woodland, open pine woods and in grassland, usually on limestone or clayey soils; sea level to 3,000 m.

FLOWERING. April to June.

DISTINGUISHING FEATURES. Leaves green or spotted or streaked with dark purple-brown. Spike cylindrical in outline. Flowers purple with the dorsal sepal and petals forming an open hood, the lateral sepals erect or suberect like rabbit's ears and obtuse to acute. Lip 3-lobed with spreading lobes, usually purple-dotted on its basal to central part, not longitudinally vaulted; spur clavate, as long as or slightly longer than the ovary, weakly upcurved.

OPPOSITE, LEFT TO RIGHT FROM TOP
France, Aveyron 18.05.2004
France, Aveyron 23.05.1996
Germany, Black Forest 25.04.1999
Germany, Stühlingen 28.04.1998
England, Surrey 26.04.2024 (25)
Germany, Black Forest 27.04.1999
Germany, Black Forest 25.04.1996
France, Aveyron 17.05.2006
France, Aveyron 20.05.2002

BELOW, LEFT TO RIGHT
France, Aveyron 21.05.2013
France, Cantal 21.05.2013

b. *Orchis mascula* subsp. *speciosa* (Mutel) Hegi, *Ill. Fl. Mitt.-Eur.*: 347 (1909). **Showy early-purple orchid**

SYNONYMS. *Orchis ovalis* F.W.Schmidt; *O. mascula* var. *speciosa* Mutel

DISTRIBUTION. C and SE Europe, from France across to Bulgaria and Poland in the NE.

HABITAT. Woodland, forest, montane meadows, on fresh and base-rich to lime-free soils; sea level to 2,600 m.

FLOWERING. May and June.

DISTINGUISHING FEATURES. 20–70 cm tall. Similar to the typical subspecies but differs in having acuminate lateral sepals that recurve at the tips. Besides, the leaves are never marked with spots, but can be finely streaked with dark purple-brown.

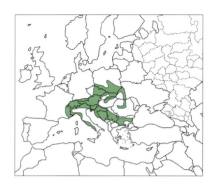

c. *Orchis mascula* subsp. *laxifloriformis* Rivas Goday & B.Rodr., *Anales Jard. Bot. Madrid* 6: 190 (1946). **Spanish early-purple orchid**

SYNONYMS. *Orchis masculolaxiflora* Lange; *O. langei* Lange ex K.Richt.

DISTRIBUTION. Spain, Portugal and SW France; also in N Morocco.

HABITAT. Woodland and forest edges, garigue, on dry, neutral, acidic and slatey soil; sea level to 1,500 m.

FLOWERING. April to June.

DISTINGUISHING FEATURES. 20–60 cm tall. Similar to the typical subspecies but differs in having a lip that is longitudinally vaulted, with no or only weak markings, and a spur that is shorter than the ovary.

d. *Orchis mascula* subsp. *scopulorum*
(Summerh.) H.Sund. ex H.Kretzschmar, Eccarius & Dietr., *Orchid Gen*. Anacamptis, Orchis, Neotinea: 308 (2007). **Madeiran early-purple orchid**

SYNONYM. *Orchis scopulorum* Summerh.

DISTRIBUTION. Endemic to Madeira.

HABITAT. Sunny and sometimes damp places in scrub, on steep basalt rocks and screes; 800–1,800 m.

FLOWERING. May and June.

DISTINGUISHING FEATURES. 25–70 cm tall. Leaves green, lacking any spots. Inflorescence tall, the spike pyramidal in outline. Flowers pale purple-pink with lateral sepals that protrude but do not recurve, and a 15–30 mm long purple-dotted lip with widely spreading lobes; spur shorter than the ovary.

LEFT TO RIGHT, FROM TOP
subsp. *speciosa*
Switzerland, Grisons 19.06.2008
Switzerland, Grisons 19.06.2008
Switzerland, Grisons 19.06.2008
subsp. *laxifloriformis*
Spain 25.05.2002 (14)
Spain 25.05.2002 (14)
Spain 25.05.2002 (14)
subsp. *scopulorum*
Madeira, 21.05. 1975 (63)
Madeira 18.05.1995 (64)

Sardinia 28.04.1990 (25)   Sardinia 28.04.1990 (25)   Sardinia 28.04.1990 (25)

e. Orchis mascula subsp. *ichnusae* Corrias, *Boll. Soc. Sarda Sci. Nat.* 21: 403 (1982).
Sardinian early-purple orchid
SYNONYMS. *Orchis ichnusae* (Corrias) J.Dev.-Tersch. & P.Devill.; *O. mascula* var. *olivetorum* Martelli
DISTRIBUTION. Endemic to Sardinia.
HABITAT. In grassy areas in maquis, bush and stony places on the edge of forest; 500–1,200 m.
FLOWERING. Mid-April to May.
DISTINGUISHING FEATURES. 15–30 cm tall. Similar to the typical subspecies but differs in having a straight (not upcurved) spur. Besides, the leaves are consistently unspotted and the flowers generally less intensely coloured.

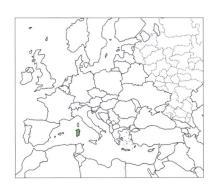

28. NEOTINEA Rchb.f., *De Pollin. Orchid.* 18: 29 (1852).

A genus of only four species that are all found in Europe. From the Mediterranean region and Europe north to Estonia, the generic range extends eastwards to the Ural Mountains and the Caspian Sea.

The 5–8 bluish green leaves can form a basal rosette or can be distributed along the stem. The long, pointed bracts are membranous and appressed to the ovary. The inflorescence is ovoid to longly cylindrical and many-flowered. Although the individual species are easily recognisable, *Neotinea* is difficult to distinguish from *Orchis* at the generic level. However, the consistent combination of small flowers and a dense inflorescence separates *Neotinea* from *Orchis* fairly well. In grassland, open woodland and garrigue, on dry, calcareous soils. Identification key on p. 429.

Neotinea maculata (Desf.) Stearn, *Ann. Mus. Goulandris* 2: 79 (1975). **Dense-flowered orchid**
SYNONYMS. *Satyrium maculatum* Desf.; *Aceras maculatum* (Desf.) Gren.; *Neotinea intacta* Rchb.f.
DISTRIBUTION. Mediterranean region, with outposts in the Canary Islands, Madeira, Brittany, Ireland and Isle of Man.
HABITAT. Grassland, scrub, coniferous and laurel forest, and tree heath scrub; sea level to 2,000 m.
FLOWERING. March to May.
DISTINGUISHING FEATURES. 10–30 cm tall. Leaves usually mottled or with well-defined purple-brown spots above and often tinted red beneath, less often uniformly dull green. Flowers tiny, creamy white to pink flushed, with connivent sepals and petals, a 1–2 mm long spur to the lip and an erect ovary; self-pollinating.

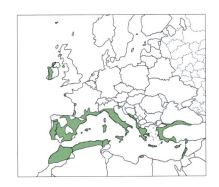

OPPOSITE, TOP LEFT TO RIGHT
Greece, Karpathos 20.03.2001 (14)
Cyprus, Troodos 16.05.1999
France, Aveyron 20.05.2009
France, Aveyron 20.05.2009
Crete 12.04.014
Crete 12.04.2014
Cyprus, Troodos 14.03.1999
Cyprus, Troodos 17.03.1999

Neotinea ustulata (L.) R.M.Bateman, Pridgeon & M.W.Chase. **Burnt-tipped orchid**
SYNONYMS. *Orchis ustulata* L.; *Odontorchis ustulata* (L.) D.Tyteca & E.Klein

a. var. *ustulata*

DISTRIBUTION. Widespread in Europe north to England, Denmark, S Sweden and Estonia; also in the Caucasus and W Siberia.
HABITAT. Sparse grassland and open oak forest, on calcareous to slightly acidic soil; sea level to 2,750 m.
FLOWERING. April to early June, but up to July in montane habitats.
DISTINGUISHING FEATURES. 5–30 cm tall. The species is readily identified by its small dark blackish-red buds in a first conical, later cylindrical, inflorescence of many small flowers that open with a dark blackish-purple sepal hood and a white lip that is spotted with red. The typical variety is a short plant and has apically incurved sepals.

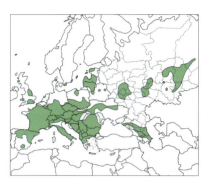

OPPOSITE, LEFT TO RIGHT
var. *ustulata*
France, Aveyron 20.05.1996
France, Aveyron 20.05.1994
France, Aveyron 18.05.1991
France, Aveyron 20.05.1994
France, Aveyron 20.05.2000
France, Aveyron 27.05.2003

b. *Neotinea ustulata* var. *aestivalis* (Kümpel) Tali, M.F.Fay & R.M.Bateman.

SYNONYMS. *Orchis ustulata* var. *aestivalis* Kümpel; *Neotinea ustulata* f. *aestivalis* (Kümpel) P.Delforge
DISTRIBUTION. From England across C Europe and the Balkans to Greece and Romania, and with northern outposts in Denmark and Estonia.
HABITAT. Dry and semi-dry grassland, sparse meadows and montane meadows, rarely in coniferous and deciduous forests; sea level to 2,000 m.
FLOWERING. Late June to August.
DISTINGUISHING FEATURES. 30–80 cm tall. Differs from the typical variety in being a taller plant with apically slightly recurved sepals. It also flowers much later than the typical variety when at the same latitude and altitude.

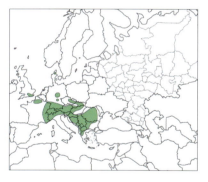

BOTTOM ROW, LEFT TO RIGHT
var. *aestivalis*
Switzerland, Aargau 05.07.2008
Switzerland, Aargau 05.07.2001
Switzerland, Aargau 30.06.2005
Switzerland, Aargau 03.07.2006

Neotinea lactea (Poir.), R.M.Bateman, Pridgeon & M.W.Chase, *Lindleyana* 12 (3): 122 (1997).

Milky orchid

SYNONYMS. *Orchis lactea* Poir.; *Odontorchis lactea* (Poir.) D.Tyteca & E.Klein; *Neotinea lactea* var. *corsica* (Viv.) P.Delforge

DISTRIBUTION. Mediterranean region from SW France and N Algeria across to W Turkey.

HABITAT. Open forest, scrub, garigue, poor grassland; sea level to 1,800 m.

FLOWERING. February to April.

DISTINGUISHING FEATURES. 7–20 cm tall, shorter but more robust than *N. tridentata*. Leaves broader than those of *N. tridentata*, silvery green, unspotted. Inflorescence cylindrical. Flowers with a white to pale green hood, veined green or red, formed by the strongly pointed sepals. Lip 8–11 mm long, with slightly recurved lobes, usually pink- or purple-spotted all over the ventral surface.

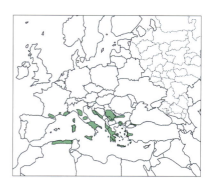

OPPOSITE PAGE, TOP LEFT TO RIGHT
Crete 05.04.2015
Italy, Sicily 10.04.2022
Crete 12.04.2015
Sardinia 14.04.2017
Sardinia 14.04.2017
Sardinia 14.04.2017
Italy, Tuscany 12.04.2006
Italy, Sicily 08.04.2022
Crete 12.04.2015

THIS PAGE, LEFT TO RIGHT
Italy, Tuscany 12.04.2006
Crete 12.04.2015
Crete 12.04.2015 (no resupination)
Crete 12.04.2015
Crete 12.04.2015

Neotinea tridentata (Scop.) R.M.Bateman, Pridgeon & M.W.Chase, *Lindleyana* 12 (3): 122 (1997).
Toothed orchid
SYNONYMS. *Orchis tridentata* Scop.; *Odontorchis tridentata* (Scop.) D.Tyteca & E.Klein

a. subsp. *tridentata*

SYNONYMS. *Neotinea commutata* (Tod.) R.M.Bateman; *N. commutata* var. *angelica* (A.Alibertis) P.Delforge

DISTRIBUTION. C and S Europe and larger Mediterranean islands (except Majorca), Turkey across to Azerbaijan and Israel.

HABITAT. Sparse grassland, open woodland and garigue, on base-rich soil; sea level to 2,200 m.

FLOWERING. April to June.

DISTINGUISHING FEATURES. 15–40 cm tall. Similar to *N. lactea* but differs in being generally taller and in having a denser, hemispherical to ovoid inflorescence and flowers with a pink to light-red hood and a lip with more or less incurved lobes.

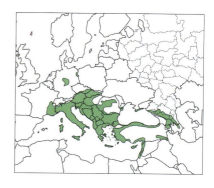

b. *Neotinea tridentata* subsp. *conica* (Willd.) R.M.Bateman, Pridgeon & M.W.Chase, *Lindleyana* 12 (3): 122 (1997).

SYNONYMS. *Orchis conica* Willd.; *O. lactea* var. *conica* (Willd.) H.Baumann & R.Lorenz

DISTRIBUTION. SW France, Iberian peninsula, Balearic Islands and NW Africa.

HABITAT. Open forest, scrub, garigue and nutrient-poor grassland; sea level to 1,000 m.

FLOWERING. February to April.

DISTINGUISHING FEATURES. 5–30 cm tall. Differs from the typical subspecies in having a conical to cylindrical, rather than hemispherical, inflorescence and smaller flowers (lip 5–7 mm long vs. 8–12 mm in subsp. *tridentata*).

OPPOSITE, TOP LEFT TO RIGHT
subsp. *tridentata*
Italy, Tuscany 11.05.1997
Greece, Fokida 30.04.1986 (11)
France, Haute Provence 10.05.2003
Crete 04.4.2016 (45)
France, Haute Provence 18.05.1997
France, Haute Provence 10.05.2003
subsp. *conica*
Portugal, Algarve 18.03.2005
Portugal, Algarve 18.03.2005
Portugal, Algarve 18.03.2005

29. OPHRYS L., *Sp. Pl.*: 948 (1753).

SYNONYMS. *Arachnites* F.W.Schmidt; *Myodium* Salisb.

DISTRIBUTION. Macaronesia, Europe across to the Caucasus, Mediterranean across to S Turkmenistan. 22 species and five partly stabilised hybrid complexes; readily recognised by the insect-like lip of the flowers. The lip possesses some special structures that are often important for identification. Thus, in most species, the basal part of the lip produces a pair of low to very prominent protuberances; in species with a 3-lobed lip, it is often the side lobes that are converted into protuberances, whereas the protuberances are more often isolated from the lip margin in species with an entire lip. Apically, the lip can simply be rounded to retuse, but in most species, it ends in a distinctly set-off triangular point or a more prominent appendage. In all *Ophrys* taxa but one, the lip is adorned by simple to complicated markings, referred to as the 'speculum'. Finally, most species possess a pair of lateral, eye-like knobs at the junction between lip and column. Morphological variation in this genus is highly complex, and strongly differing views exist as to how many species and infraspecific taxa should be recognised. The situation is further complicated by frequent hybridisation and because the hybrids are often fully fertile. It seems that, in the most extreme cases, recurrent hybridisation and back-crossing to the parental taxa have created not only confusing local hybrid swarms but even partly stabilised, species-like hybrid complexes, which sometimes extend into geographic regions where both parental taxa are missing. We treat such putative, species-like hybrid complexes under binary hybrid names at the end of the *Ophrys* account. Most of the complexes are strongly morphologically heterogenous, probably because more than one subspecies of each parental species are often involved. Some complexes contain fairly well-defined forms, as well as populations that cannot be referred to any such form readily. Consequently, we have chosen simply to refer to the distinct forms as informal hybrid forms, so-called 'nothomorphs' (nm.). Identification key on p. 430.

Ophrys insectifera L., *Sp. Pl.*: 948 (1753). **Fly orchid**
SYNONYM. *Orchis insectifera* (L.) Crantz

a. subsp. *insectifera*

SYNONYMS. *Ophrys insectifera* var. *myodes* L.; *O. muscifera* Huds.

DISTRIBUTION. N and C Europe south to N Spain, S Italy, N Greece and Romania; absent from Iceland, Scotland and much of E Europe.

HABITAT. Woodland, wooded meadows, grassland and calcareous fens; up to 1,700 m.

FLOWERING. May to July.

DISTINGUISHING FEATURES. 15–50 cm tall. A distinctive taxon with a lax inflorescence and small flowers with green sepals and slender, linear, dark brown petals that are longer than the column. The lip is longer than broad, 3-lobed with brown (occasionally yellow-tipped) lobes and a central shiny bluish speculum; side lobes narrow and spreading, mid-lobe much larger, emarginate. Column with reddish anther locules.

OPPOSITE, TOP LEFT TO RIGHT
France, Aveyron 26.05.2003
France, Aveyron 20.05.1996
France, Haute Provence 10.05.2001
Switzerland, Basel-Land 12.06.2005
Switzerland, Basel-Land 05.06.2005
France, Cote d'Azure 22.05.2007
Switzerland, Basel-Land 15.06.2007
France, Aveyron 21.05.1971
France, Aveyron 20.05.2006
France, Aveyron 22.05.2003
England, Hampshire 12.05.2009 (25)

b. *Ophrys insectifera* subsp. *aymoninii* Breistr., *Bull. Soc. Bot. France, Lett. Bot.* 128: 71 (1981).

SYNONYM. *Ophrys aymoninii* (Breistr.) Buttler

DISTRIBUTION. Endemic to CS France.

HABITAT. Open beech and pine forests and grassland; 500–900 m.

FLOWERING. May and June.

DISTINGUISHING FEATURES. 15–60 cm tall. Differs from the other two subspecies in having greenish petals (sometimes weakly suffused with brown) that are longer than the column, a lip that is about as broad as long, with broad yellow margins, and a column with yellow anther locules.

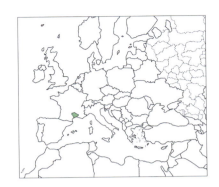

c. *Ophrys insectifera* subsp. *subinsectifera* (C.E.Hermos. & Sabando) O.Bolós & Vigo, *Fl. Man. Països Catalans* 4: 661 (2001).

SYNONYM. *Ophrys subinsectifera* C.E.Hermos. & Sabando

DISTRIBUTION. Endemic to NE Spain.

HABITAT. In full sun to light shade in grassland, scrub and open broadleaf forest; 400–1,100 m.

FLOWERING. May and June.

DISTINGUISHING FEATURES. Small-flowered. Mainly distinguished from the other two subspecies by its petals being shorter than or as long as the column. Its shallowly 3-lobed lip with yellow-tipped lobes is also characteristic.

OPPOSITE, TOP LEFT TO RIGHT
subsp. *aymoninii*
France, Aveyron 13.05.2007
France, Aveyron 18.05.1999
France, Aveyron 20.05.2003
France, Aveyron 18.05.2013
France, Aveyron 18.05.2005
France, Aveyron 26.05.1997
France, Aveyron 26.05.1997
France, Aveyron 18.05.2001
France, Aveyron 20.05.2006 (lip entire)
France, Aveyron 19.05.2009
France, Aveyron 18.05.2001

BELOW, ALL IMAGES
subsp. *subinsectifera*
Spain, Catalunya 29.05.2000 (2)

Ophrys speculum Link, *J. Bot. (Schrader)* 1799 (2): 324 (1800), nom. cons. **Mirror of Venus orchid**

a. subsp. *speculum*

SYNONYMS. *Ophrys ciliata* Biv.; *O. vernixia* subsp. *ciliata* (Biv.) Del Prete; *O. vernixia* subsp. *orientalis* Paulus

DISTRIBUTION. Widespread in the Mediterranean from Portugal and NW Africa across to Turkey and N Syria; absent from Cyprus and most of the Balkans.

HABITAT. In full sun to light shade in garigue, poor grassland, roadside verges, open pine woods, fallow fields and olive groves; sea level to 1,200 m.

FLOWERING. February to April.

DISTINGUISHING FEATURES. 5–30 cm tall. The species is readily recognised by its distinctive lip which has a large blue convex speculum bordered by a fringe of red-brown hairs. (*Ophrys bertolonii* also has a relatively large blue speculum but the blackish lip is saddle-shaped.) The typical subspecies has dark green sepals longitudinally streaked with brown and a lip with lobes with spreading to only slightly recurved margins; side lobes obliquely oblong, 0.8–1.5 times as long as wide; mid-lobe 0.6–1.1 times as long as wide.

LEFT TO RIGHT FROM TOP
subsp. *speculum*
Portugal, Algarve 13.03.2001
Majorca 18.04.2013
Majorca 18.04.2013
Portugal, Coimbra 13.04.1992
Majorca 18.04.2013
Majorca 18.04.2013
Majorca 18.04.2013
Italy, Sicily 16.04.2022

b. ***Ophrys speculum* subsp. *lusitanica*** O. & E.Danesch, *Orchidee (Hamburg)* 20: 21 (1969).

SYNONYMS. *Ophrys vernixia* Brot.; *O. vernixia* subsp. *lusitanica* (O. & E.Danesch) H.Baumann & Künkele; *O. lusitanica* (O. & E.Danesch) Paulus & Gack; *O. ciliata* subsp. *lusitanica* (O. & E.Danesch) H.Baumann, Künkele & R.Lorenz

DISTRIBUTION. WC & S Portugal and S Spain.

HABITAT. Full sun to light shade in poor grassland and garigue on calcareous soil; sea level to 500 m.

FLOWERING. March to May.

DISTINGUISHING FEATURES. 15–50 cm tall. Differs from the other two subspecies by the following combination of features: sepals pale green, sometimes weakly suffused with brown; lip with lobes with distinctly recurved margins; side lobes obliquely (linear-)oblong, more than 1.5 times as long as wide; mid-lobe 1–2.2 times as long as wide.

LEFT TO RIGHT
subsp. *lusitanica*
Portugal, Coimbra 14.04.1992
Portugal, Coimbra 14.04.1992
Portugal, Coimbra 14.04.1992
Portugal, Coimbra 14.04.1992
Portugal, Coimbra 14.04.1992

c. *Ophrys speculum* subsp. *regis-ferdinandii*
(Acht. & Kellerer ex Renz) Soó, *Acta Bot. Acad. Sci. Hung.* 16: 390 (1970).

SYNONYMS. *Ophrys speculum* f. *regis-ferdinandii* Acht. & Kellerer ex Renz; *O. regis-ferdinandii* (Acht. & Kellerer ex Renz) Buttler

DISTRIBUTION. E Aegean Islands to SW Turkey.

HABITAT. Full sun to light shade in poor grassland, garigue, maquis and open forest; sea level to 400 m.

FLOWERING. March to April.

DISTINGUISHING FEATURES. 5–30 cm tall. Differs from the other two subspecies in having the following combination of features: sepals green, strongly suffused with brown; lip with lobes with distinctly recurved margins; side lobes obliquely linear-oblong, more than 1.5 times as long as wide; mid-lobe 2–4 times as long as wide.

OPPOSITE, LEFT TO RIGHT
subsp. *regis-ferdinandii*
Rhodes 06.04.1966 (10)
Rhodes 01.04.1988 (20)
Rhodes 04.1991 (22)
Turkey, Aeydin 07.04.1970 (10)
Rhodes 31.03.1988 (20)
Rhodes 06.04.1989 (13)

Ophrys bombyliflora Link, *J. Bot. (Schrader)* 1799 (2): 325 (1800). **Bumble-bee orchid**

SYNONYM. *Arachnites bombyliflorus* (Link) Tod.

a. var. *bombyliflora*

DISTRIBUTION. Mediterranean region from Canary Islands to NE Libya and W Turkey.

HABITAT. In full sun to light shade in garigue, grassland, olive groves, open forest and moist meadows; sea level to 900 m.

FLOWERING. February to May.

DISTINGUISHING FEATURES. 5–20 cm, rarely more, tall; often forming dense colonies (clones). The species is easily recognisable from its condensed inflorescence of mainly green flowers and its vaulted, deeply 3-lobed brown lip with a blurred speculum. The lip of typical variety is at least 2.5 times as long as the stigmatic cavity; mid-lobe obovate, 6–8 mm wide; speculum obscure, dull (violet-)grey.

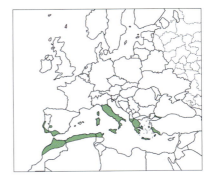

OPPOSITE, LEFT TO RIGHT
var. *bombyliflora*
Portugal, Algarve 18.04.2005
Italy, Tuscany 18.04.1998
Italy, Tuscany 26.04.2002
Italy, Tuscany 28.04.1996
Italy, Tuscany 18.04.2009
Crete 02.04.2014
Crete 02.04.2014
Crete 03.04.2014
Portugal, Algarve 18.04.2005
Crete 02.04.2014
Crete 02.04.2014

b. *Ophrys bombyliflora* var. *parviflora* (Mifsud)
Mifsud, *J. Eur. Orch.* 46: 684 (2014).

SYNONYM. *Ophrys bombyliflora* f. *parviflora* Mifsud

DISTRIBUTION. Originally described from Malta and later in Sicily and Sardinia. A few records from other regions are in need of verification.

HABITAT. Lowland garigue.

FLOWERING. Late February to early April.

DISTINGUISHING FEATURES. Differs from the typical variety by the following features: lip 1.5–2 times as long as the stigmatic cavity; mid-lobe oblong, 4–5 mm wide; speculum usually distinct, milky-blue to almost white (rarely obscure).

BELOW, LEFT TO RIGHTS
var. *parviflora*
Italy, Sicily 07.04.2018
Italy, Sicily 07.04.2018
Italy, Sicily 09.04.2018

OPPOSITE, LEFT TO RIGHT
O. tenthredinifera
Crete 10.04.2014
Sardinia 10.04.2017
Sardinia 15.04.2017
Italy, Puglia 10.04.2016
Crete 08.04.2015
Italy, Puglia 10.04.2016

Ophrys tenthredinifera Willd., *Sp. Pl.* 4: 67 (1805).
Sawfly orchid

SYNONYMS. *Arachnites tenthredinifer* (Willd.) Tod.; *Ophrys tenthredinifera* subsp. *villosa* (Desf.) H.Baumann & Künkele; *O. tenthredinifera* subsp. *guimaraesii* D.Tyteca; *O. aprilia* Devillers & Devillers-Tersch.; *O. dictynnae* P.Delforge; *O. leochroma* P.Delforge; *O. ulyssea* P.Delforge; *O. tenthredinifera* subsp. *spectabilis* Kreutz & Zelesny; *O. spectabilis* (Kreutz & Zelesny) Paulus; *O. korae* M.Hirth & Paulus; *O. riphaea* (F.M.Vázquez) P.Delforge; *O. amphidami* P.Delforge; *O. lycomedis* P.Delforge

DISTRIBUTION. Mediterranean region from Portugal and NW Africa across to Turkey but extinct in Corsica and Cyprus.

HABITAT. In full sun to light shade in grassland, rocky places, garigue, roadside verges, open pine woods and olive groves; sea level to 1,800 m.

FLOWERING. February to May.

DISTINGUISHING FEATURES. 10–30 cm, or more, tall. A distinctive orchid with rose-purple to white sepals and petals, a rounded column and a large (sub)entire lip with a broad yellow hairy margin, recurved sides and a small speculum. A prominent tuft of hairs close to the tip of the lip is particularly characteristic.

NOTE. Two Italian taxa should be commented upon, viz. *O.* ×*normanii* J.J.Wood from S Sardinia and *O.* ×*tardans* O.Danesch & E.Danesch from S Puglia. When first described, the former was interpreted as the hybrid *O. holosericea* subsp. *chestermanii* × *O. tenthredinifera*, and the latter as *O.holosericea* subsp. *candica* × *O. tenthredinifera*. Indeed, both are morphologically intermediate between the parental taxa proposed, but further studies of their taxonomic position are desirable.

BELOW, LEFT TO RIGHT
O. tenthredinifera
Italy, Sicily 08.04.2022
Italy, Puglia 17.04.2016
Portugal, Coimbra 10.04.1992
Italy, Sicily 08.04.2022
Italy, Puglia 17.04.16
Italy, Puglia 14.04.2018

Ophrys lutea Cav., *Icon.* 2: 46 (1793).
Yellow bee orchid

SYNONYM. *Arachnites luteus* (Cav.) Tod., *Orchid. Sicul.*: 95 (1842).

a. subsp. *lutea*

SYNONYMS. *Ophrys phryganae* Devillers-Tersch. & Devillers; *O. corsica* Soleirol ex G. & W.Foelsche; ?*O. hellenica* Devillers & Devillers-Tersch.; *O. penelopeae* Paulus; *O.* ×*sulphurea* Gennaio & Medagli

DISTRIBUTION. The Mediterranean region, particularly in the Iberian Peninsula and NW Africa, across to SW Turkey; in W France extending north to Brittany.

HABITAT. Grasslands, garigue, maquis, open pine and oak woods, olive groves and roadsides in full sun to light shade, usually on calcareous soils; sea level to 1,800 m.

FLOWERING. February to May.

DISTINGUISHING FEATURES. 10–40 cm tall. Dorsal sepal parallel to column; petals glabrous. Lip lightly vaulted, 13–19 mm broad with a distinct knee-like bend at the base, lacking a terminal appendage but provided with spreading or upcurved, usually bright yellow, margins that are usually 3–6 mm broad. Column rounded.

BELOW, LEFT TO RIGHT
France, Aveyron, 22.05.1999
France, Aveyron 20.05.2004
France, Aveyron 18.05.2007
Italy, Sicily 12.04.2018
Italy, Sicily 07.04.2018
Italy, Sicily 07.04.2018
Italy, Sicily 07.042018
Italy, Sicily 07.04.2018

b. *Ophrys lutea* subsp. *aspea* (Devillers-Tersch. & Devillers) Faurh., *J. Eur. Orch.* 41: 194 (2009).

SYNONYMS. *Ophrys aspea* Devillers-Tersch. & Devillers; *O. subfusca* subsp. *aspea* (Devillers-Tersch. & Devillers) Kreutz

DISTRIBUTION. N Tunisia and NW Libya.

HABITAT. In light shade in forest dominated by oaks, pines or Phoenicean junipers; occasionally under shrubs in garigue; sea level to 400 m.

FLOWERING. March and April.

DISTINGUISHING FEATURES. 10–25 cm tall. Differs from the other three subspecies in having the following combination of features: lip strongly vaulted, 8–12 mm broad, with a distinct knee-like bend at the base and yellow margins that are 1–3 mm broad.

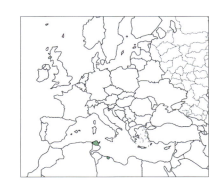

c. *Ophrys lutea* subsp. *galilaea* (H.Fleischm. & Bornm.) Soó, *Notizbl. Bot. Gart. Berlin-Dahlem* 9: 906 (1926).

SYNONYMS. *Ophrys sicula* Tineo; *O. lutea* var. *minor* (Tod.) Guss.; *O. galilaea* H.Fleischm. & Bornm.; *O. archimedea* P.Delforge & M.Walravens; *O. numida* Devillers-Tersch. & Devillers; *O. lepida* S. & J.-M. Moingeon; *O. lutea* subsp. *quarteirae* Kreutz, M.R.Lowe & Wucherpf.; *O. pseudomelena* Turco, Medagli & D'Emerico; ?*O. cythnia* P.Delforge & Onckelinx

DISTRIBUTION. Widespread in the Mediterranean from Portugal and NW Africa to Israel. Absent from France and very rare in the Iberian Peninsula.

HABITAT. As for the typical subspecies.

FLOWERING. January to June.

DISTINGUISHING FEATURES. 5–40 cm tall. Differs from the other three subspecies in having the following combination of features: lip lightly vaulted, 4–12 mm broad, porrect and with a yellow margin that is usually 2–3 mm broad.

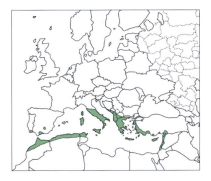

OPPOSITE, LEFT TO RIGHT FROM TOP

subsp. *aspea*
Tunisia, Nabeul 01.04.2004 (2)
Tunisia, Nabeul 01.04.2004 (2)
Tunisia, Nabeul 01.04.2004 (2)

subsp. *galilaea*
Cyprus 18.03.1995
Crete 05.04.2014
Italy, Sicily 08.04.2022
Cyprus 15.03.1998
Cyprus, Akamas 13.03.2005
Italy, Sicily 08.04.2022
Italy, Sicily 09.04.2018

d. *Ophrys lutea* subsp. *melena* Renz, *Repert. Spec. Nov. Regni Veg.* 25: 264 (1928).

SYNONYMS. *Ophrys galilaea* subsp. *melena* (Renz) Del Prete; *O. melena* (Renz) Paulus & Gack; *O. praemelena* S.Hertel & Presser

DISTRIBUTION. Mainly SW Greece but also in S Albania, Crete and a few Aegean islands.

HABITAT. Full sun to light shade in garigue, maquis and abandoned wine terraces on calcareous soil; sea level to 1,300 m.

FLOWERING. March and April.

DISTINGUISHING FEATURES. 10–40 cm tall. Differs from the other three subspecies in having the following combination of features: lip lightly vaulted, 7–12 mm broad, porrect and with a narrow yellow margin that is up to 1 mm broad (sometimes absent).

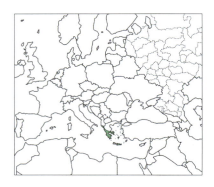

Ophrys fusca Link, *J. Bot. (Schrader)* 1799 (2): 324 (1800). **Dingy bee orchid**

SYNONYM. *Arachnites fuscus* (Link) Tod.

a. subsp. *fusca*

SYNONYMS. *Ophrys funerea* Viv.; *O. lutea* subsp. *funerea* (Viv.) Batt.; *O. obaesa* Lojac.; *O. leucadica* Renz; *O. punctulata* Renz; *O. ficuzzana* H.Baumann & Künkele; *O. attaviria* D. & U.Rückbr., Wenker & S.Wenker; *O. mesaritica* Paulus, C. & A.Alibertis; *O. calocaerina* Devillers-Tersch. & Devillers; *O. lupercalis* Devillers-Tersch. & Devillers; *O. sulcata* Devillers-Tersch. & Devillers; *O. zonata* Devillers-Tersch. & Devillers; *O. lojaconoi* P.Delforge; *O. parosica* P.Delforge; *O. phaseliana* D. & U.Rückbr.; *O. creberrima* Paulus; *O. cressa* Paulus; *O. creticola* Paulus; *O. marmorata* G. & W.Foelsche; *O. thriptiensis* Paulus; *O. arnoldii* P.Delforge; *O. lucentina* P.Delforge; *O. caesiella* P.Delforge; *O. flammeola* P.Delforge; *O. gazella* Devillers-Tersch. & Devillers; *O. hespera* Devillers-Tersch. & Devillers; *O. lucana* P.Delforge, Devillers-Tersch. & Devillers; *O. lucifera* Devillers-Tersch. & Devillers; *O. peraiolae* G. & W.Foelsche, M. & O.Gerbaud; *O. proxima* C.E.Hermos., Benito & Soca; *O. africana* G. & W.Foelsche; *O. dianica* M.R.Lowe, J.Piera, M.B.Crespo & J.E.Arnold; *O. eptapigiensis* Paulus; *O. lindia* Paulus; *O. parvula* Paulus; *O. ortuabis* M.P.Grasso & Manca;

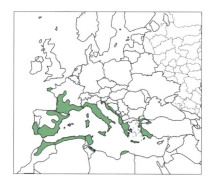

OPPOSITE, LEFT TO RIGHT
Italy, Puglia
France, Aveyron 17.05.1989
France, Aveyron 18.05.1989
France, Aveyron 23.05.2001
France, Aveyron 17.05.2006
Italy, Sicily 11.04.2022

OPHRYS | 229

Greece, Peloponnese 06.04.1994 (20)   Greece, Peloponnese 06.04.1994 (20)   Greece, Peloponnese 25.04.2000 (2)

O. cesmeensis (Kreutz) P.Delforge; *O. fabrella* Paulus & Ayasse ex P.Delforge; *O. gackiae* P.Delforge; *O. perpusilla* Devillers-Tersch. & Devillers; *O. sabulosa* Paulus & Gack ex P.Delforge; *O. clara* F.M.Vázquez & S.Ramos; *O. delforgei* Devillers-Tersch. & Devillers; *O. kedra* Paulus; *O. pallidula* Paulus; *O. phaidra* Paulus; *O. rueckbrodtiana* W.Hahn; *O. meropes* P.Delforge; *O. theophrasti* Devillers & Devillers-Tersch.; *O. thracica* (Kreutz) Devillers & Devillers-Tersch.; *O. malacitana* M.R.Lowe, I.Phillips & Paulus; *O. achillis* P.Delforge

DISTRIBUTION. Throughout the Mediterranean from Portugal and NW Africa to Turkey; also in W France.

HABITAT. Grassland, garigue, maquis, olive groves, open oak and pine woods, rocky grassland and roadside verges, in full sun to light shade; sea level to 1,500 m.

FLOWERING. January to June, with a peak in March and April.

DISTINGUISHING FEATURES. 10–40 cm tall. Similar to *O. lutea*, but with a lip with recurved margins, and to *O. omegaifera*, but with a longitudinal furrow at the base of the lip. Lip 7–23 mm long, green underneath, lightly to strongly downcurved for most of its length, distinctly vaulted in cross-section, with or without a narrow yellow margin; side lobes more or less downcurved; speculum dull, occasionally ending in a more or less obscure omega-shaped mark.

OPPOSITE, LEFT TO RIGHT
subsp. *fusca*
Crete 10.04.2014
Italy, Sicily 11.04.2022
Sardinia 10.04.2017
Spain, Navarre 13.05.2014 (45)
Italy, Sicily 11.04.2022
Italy, Sicily 12.04.2022
subsp. *blitopertha*
Rhodes 28.03.2000 (56)
Rhodes 28.03.2000 (56)
Samos 04.04.1991 (1)

b. Ophrys fusca subsp. *blitopertha* (Paulus & Gack) Faurh. & H.A.Pedersen, *Orchidee (Hamburg)* 53: 345 (2002).

SYNONYMS. *Ophrys blitopertha* Paulus & Gack; *O. persephonae* Paulus; *O. subfusca* subsp. *blitopertha* (Paulus & Gack) Kreutz; ?*O. urteae* Paulus

DISTRIBUTION. Aegean Islands to SW Turkey.

HABITAT. Garigue and old olive groves; sea level to 400 m.

FLOWERING. Late March to May.

DISTINGUISHING FEATURES. 7–20 cm tall. Differs from the other four subspecies in having the following combination of features: lip 8–17 mm long, green underneath, straight, nearly flat in cross-section and with a broad yellow (to yellow-brown) margin; side lobes spreading; speculum dull.

c. Ophrys fusca subsp. cinereophila (Paulus & Gack) Faurh., *Orchidee (Hamburg)* 53: 345 (2002).

SYNONYMS. *Ophrys cinereophila* Paulus & Gack; *O. subfusca* subsp. *cinereophila* (Paulus & Gack) Kreutz

DISTRIBUTION. From S Greece across SW Turkey to Cyprus and N Syria.

HABITAT. Full sun to light shade in garigue, maquis, olive groves and open pine forest, usually on calcareous soil; sea level to 700 m.

FLOWERING. February to mid-April.

DISTINGUISHING FEATURES. 7–25 cm tall. Differs from the other three subspecies in having the following combination of features: flowers arranged in a spiral on the upper third (or more) of the stem; sepals pale green; lip 7–12 mm long, green underneath, abruptly downcurved near the base, with or without a narrow yellow margin; side lobes downcurved; speculum dull.

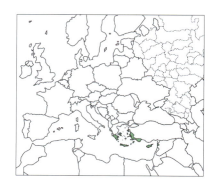

FROM TOP, LEFT TO RIGHT
subsp. *cinereophila*
Cyprus 12.03.1998
Crete 08.04.2014
Crete 02.04.2014
Cyprus 16.03.1999
Cyprus 15.03.1996
Cyprus 03.04.2014
Cyprus 13.03.1994

d. Ophrys fusca subsp. iricolor (Desf.) K.Richt., *Pl. Eur.* 1: 261 (1890).

SYNONYMS. *Ophrys iricolor* Desf.; *O. eleonorae* Devillers-Tersch. & Devillers; *O. vallesiana* Devillers-Tersch. & Devillers; *O. astypalaeica* P.Delforge; ?*O. hospitalis* P.Delforge

DISTRIBUTION. Scattered in the Mediterranean region from Portugal and NW Africa across to Turkey.

HABITAT. As for the typical subspecies; sea level to 1,100 m.

FLOWERING. February to May.

DISTINGUISHING FEATURES. 10–40 cm, or more, tall. Differs from the other four subspecies in having the following combination of features: lip 14–26 mm long, wine-red underneath, straight, vaulted in cross-section; side lobes downcurved; speculum shiny blue.

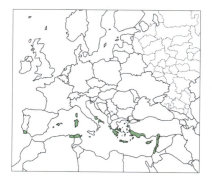

FROM THIRD ROW, LEFT TO RIGHT
subsp. *iricolor*
Cyprus 18.03.2011
Crete 07.04.2014
Crete 02.04.2014
Crete 02.04.2014
Crete 07.04.2014
Crete 03.04.2014

234 | OPHRYS

e. *Ophrys fusca* subsp. *pallida* (Raf.) E.G.Camus
in E.G.Camus & A.A.Camus, *Iconogr. Orchid. Europe*: 293 (1929).

SYNONYMS. *Ophrys pallida* Raf.; *O. pectus* Mutel

DISTRIBUTION. NW Sicily and NE Algeria.

HABITAT. Full sun to light shade in grassland, garigue, maquis and open forest on calcareous soil; sea level to 1,200 m.

FLOWERING. Late February to early May.

DISTINGUISHING FEATURES. 10–30 cm tall. Differs from the other four subspecies in having the following combination of features: flowers densely clustered in the upper quarter of the stem; sepals white or very nearly so; lip 7–11 mm long, strongly and abruptly downcurved near the base; side lobes downcurved; speculum dull.

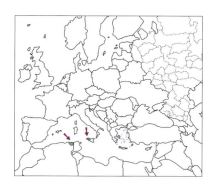

ALL IMAGES
subsp.*pallida*
Italy, Sicily 08.04.2022

Ophrys omegaifera H.Fleischm., *Oesterr. Bot. Z.* 74: 184 (1925).

SYNONYMS. *Ophrys lutea* subsp. *omegaifera* (H.Fleischm.) Soó; *O. fusca* subsp. *omegaifera* (H.Fleischm.) E.Nelson

a. subsp. *omegaifera*

SYNONYMS. *Ophrys polycratis* P.Delforge; *O. appolonae* Paulus & M.Hirth

DISTRIBUTION. Crete, Aegean Islands, SE mainland Greece and SW Turkey.

HABITAT. Full sun to light shade in garigue, olive groves and open pine forest on calcareous soil; sea level to 1,000 m.

FLOWERING. January to early May.

DISTINGUISHING FEATURES. 8–30 cm tall. Distinguished from *O. fusca* by its flat (not furrowed) base of the lip and in having a speculum that ends in a distinct omega-shaped mark that is usually bluish in newly opened flowers. The flowers are inclined-spreading and the lip abruptly downcurved in its basal part, with short-hairy lobes.

NOTE. Early-flowering Greek populations of large-flowered plants with dark ground colour of the lip may be recognised as var. *basilissa* (C. & A.Alibertis & H.R.Reinhard) Faurh. However, this form overlaps with the typical form in both flower size, lip colour and flowering time.

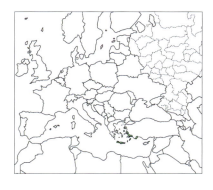

CLOCKWISE FROM TOP LEFT
Crete 07.04.2015
Crete 03.02.2014
Crete 07.04.2015
Crete 04.04.2015
Crete 02.04.2014
Crete 07.04.2015
Crete 04.04.2015

b. *Ophrys omegaifera* subsp. *dyris* (Maire) Del Prete, *Webbia* 38: 213 (1984).

SYNONYMS. *Ophrys dyris* Maire; *O. fusca* subsp. *dyris* (Maire) Soó; *O. algarvensis* D.Tyteca, Benito & M.Walravens

DISTRIBUTION. S Iberian Peninsula, Majorca and N Morocco.

HABITAT. Full sun to light shade in grassland, garigue and open conifer forest; sea level to 2,000 m.

FLOWERING. March to May.

DISTINGUISHING FEATURES. 10–35 cm, or more, tall. Differs from the other four subspecies in having the following combination of features: flowers inclined-spreading; lip abruptly downcurved in its basal part; speculum usually red-brown, pale at base, the omega-shaped mark usually white to light grey in newly opened flowers

c. *Ophrys omegaifera* subsp. *fleischmannii* (Hayek) Del Prete, *Webbia* 38: 213 (1984).

SYNONYMS. *Ophrys fleischmannii* Hayek; *O. heldreichii* H.Fleischm., nom illeg.; *O. fusca* subsp. *fleischmannii* (Hayek) Soó

DISTRIBUTION. Probably endemic to Crete.

HABITAT. Full sun to light shade in garigue, maquis, olive groves and open pine forest, usually on calcareous soil; sea level to 1,200 m.

FLOWERING. February to April.

DISTINGUISHING FEATURES. 8–20 cm tall. Differs from the other four subspecies in having the following combination of features: flowers erect; lip abruptly downcurved in its basal part, with long-hairy lobes; speculum usually bluish, not pale at base, the omega-shaped mark usually white to light grey in newly opened flowers.

LEFT TO RIGHT FROM TOP
subsp. *dyris*
Spain, Andalucia 18.04.2001 (45)
Spain, Navarre 11.05.2011 (45)
Spain, Navarre 11.05.2011 (45)
Spain, Andalucia 18.04.2001 (45)
Spain, Andalucia 18.04.2001 (45)
susbp. *fleischmannii*
Crete 02.04.2014
Crete 02.04.2014
Crete 02.04.2014
Crete 14.04.2003 (45)
Crete 14.04.2003 (45)

d. *Ophrys omegaifera* subsp. *hayekii*
(H.Fleischm. & Soó) Kreutz, *Kompend. Eur. Orchid.*: 111 (2004).
SYNONYMS. *Ophrys fusca* subsp. *hayekii* H.Fleischm. & Soó; *O. atlantica* subsp. *hayekii* (H.Fleischm. & Soó) Soó; *O. mirabilis* Geniez & Melki
DISTRIBUTION. Sicily and N Tunisia.
HABITAT. Full sun to light shade in grassland, garigue and edges of forest on calcareous soil; sea level to 900 m.
FLOWERING. Mid-April to early May.
DISTINGUISHING FEATURES. 15–22 cm tall. Differs from the other four subspecies in having the following combination of features: lateral sepals more than 1.75 times as long as wide, broadest at or below the middle; lip straight or only slightly downcurved at the base; speculum clear, unmarbled, the omega-shaped mark obscure.

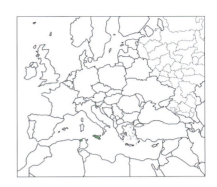

TOP ROW, LEFT TO RIGHT
subsp. *hayekii*
Sicily 03.05.2001 (3)
Sicily, Solarino 25.04.2004 (56)
Italy, Sicily 12.04.2022

e. *Ophrys omegaifera* subsp. *israelitica*
(H.Baumann & Künkele) G. & K.Morschek, *Orchids Cyprus*: 126 (1996).
SYNONYM. *Ophrys israelitica* H.Baumann & Künkele
DISTRIBUTION. C Aegean Islands, Cyprus and S Turkey to Jordan.
HABITAT. Full sun to light shade in garigue, maquis, poor grassland, olive groves and open conifer forest; sea level to 1,300 m.
FLOWERING. January to April.
DISTINGUISHING FEATURES. 8–20 cm, or more, tall. Differs from the other four subspecies in having the following combination of features: lateral sepals less than 1.75 times as long as wide; broadest at or above the middle; lip straight or only slightly downcurved at the base; speculum usually marbled, the omega-shaped mark distinct.

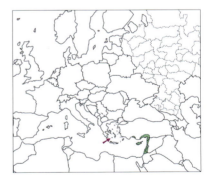

FROM SECOND ROW, LEFT TO RIGHT
susbp. *israelitica*
Cyprus 17.03.2004
Cyprus 15.03.1999
Cyprus 16.03.2001
Cyprus 11.03.2009
Cyprus 17.03.2004
Cyprus 18.03.1995
Cyprus 15.03.2011
Cyprus 15.03.2007
Cyprus 16.03.2009

Ophrys atlantica Munby, *Bull. Soc. Bot. France* 3: 108 (1856). **Atlantic bee orchid**

SYNONYMS. *Ophrys fusca* var. *durieui* Rchb.f.; *O. fusca* var. *atlantica* (Munby) Coss.; *O. fusca* subsp. *durieui* (Rchb.f.) Soó; *O. fusca* subsp. *atlantica* (Munby) E.G.Camus

DISTRIBUTION. S Spain (Andalusia), NW Africa.

HABITAT. In full sun to light shade on basic, dry to somewhat moist soils in deciduous and pine woods, garigue and scrubby grassland; sea level to 1,500 m.

FLOWERING. March to early June. Early in Spain, later in NW Africa.

DISTINGUISHING FEATURES. 10–35 cm tall. Sepals pale to yellowish green; petals recurved, olive-green or brownish green, with wavy margins. Lip 15–25 mm long, saddle-shaped, 3-lobed in the apical part, mid-lobe hardly longer than the side lobes; speculum simple, shiny blue.

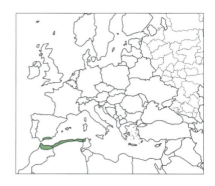

LEFT TO RIGHT
Spain 10.04.1996 (1)
Spain, Andalucia 19.04.1972 (1)
Spain, Andalucia 04.1984 (22)

Ophrys apifera Huds., *Fl. Angl.*: 340 (1762). **Bee orchid**

SYNONYMS. *Orchis apifera* (Huds.) Salisb.; *Ophrys apifera* var. *chlorantha* (Hegetschw.) Nyman

DISTRIBUTION. Mediterranean to N Iran and Europe north to the British Isles and S Scandinavia; absent from much of E Europe.

FLOWERING. April to July.

HABITAT. Grasslands, dune slacks, garigue, maquis, open deciduous and pine woods, roadsides and waste places, in full sun to light shade, especially on calcareous soils; sea level to 1,800 m.

DISTINGUISHING FEATURES. 15–50 cm tall, rarely more. Easily recognised by its prolonged, sigmoidly bent column apex, self-pollinating flowers each producing a capsule and with a bee-like lip.

NOTE. When a flower opens, the curved anther continues its growth, gradually pulling the pollinia out of the anther compartments (as their base remains attached). Because the pollinia stalks are flaccid, the extracted pollinia tilt forwards and downwards into a position from where a gust of wind easily makes them pivot onto the stigma, thus effecting self-pollination of the flower.

The spontaneous self-pollination in *O. apifera* is unique within *Ophrys*. The lack of genetic recombination and the circumstance that the

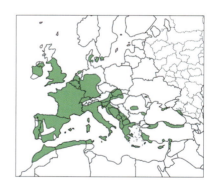

LEFT TO RIGHT
Germany, Baden-Württemberg 23.05.2017
France, Alsace 01.06.2017
France, Haute-Provence 08.05.2002
France, Aveyron 22.05.2007
France, Haute-Provence 08.05.2002
France, Alsace 09.06.2016

OPHRYS | 241

species does not depend on precise adaptations to specific pollinators are probably important reasons why aberrant (to some extent even 'monstrous') forms are much more common in *O. apifera* than in any other *Ophrys* species–there simply is no strong selection against them. More than 30 such forms have been described formally, and five are presented here.

a. forma *austroalsatica* Sprunger & Rolf Kühn, *Orchidee, Taxon. Mitt.* 8, 8: 52 (2022).

b. forma *aurita* (Moggr.) Soó, *Bot. Arch.* 3: 34 (1928).
SYNONYMS. *Ophrys insectifera* subvar. *aurita* Moggr.; *O. apifera* var. *aurita* (Moggr.) Gremli

c forma *bicolor* (E.Nelson) P.D.Sell, *Fl. Gr. Brit. Ireland* 5: 365 (1996).
SYNONYM. *Ophrys apifera* var. *bicolor* E.Nelson

d. forma *friburgensis* (Freyhold) M.Schulze, *Oesterr. Bot. Z.* 49: 269 (1899).
SYNONYMS. *Ophrys apifera* var. *friburgensis* Freyhold; *O. botteronii* Chodat; *O. apifera* var. *botteronii* (Chodat) Brand; *O. apifera* subsp. *botteronii* (Chodat) O.Nägeli; *O. friburgensis* (Freyhold) O.Nägeli; *O. apifera* f. *botteronii* (Chodat) P.D.Sell.

e. forma *trollii* (Hegetschw.) P.D.Sell, *Fl. Gr. Brit. Ireland* 5: 365 (1996). **Wasp orchid**
SYNONYMS. *Ophrys trollii* Hegetschw.; *O. apifera* var. *trollii* (Hegetschw.) Rchb.f.; *O. apifera* subsp. *trollii* (Hegetschw.) K.Richt.
NOTE. Considered to be a temperature-induced form.

OPPOSITE PAGE, LEFT TO RIGHT
O. apifera
France, Alsace 16.06.2019
France, Alsace 16.06.2019
France, Provence 12.05.2009
France, Aveyron 24.03.1998
France, Aveyron 24.03.1998
Italy, Umbria 20.05.2008 (21)
forma *austroalsatica*
France, Alsace 28.05.2022
France, Alsace 06.06.2021
France, Alsace 06.06.2021
forma *bicolor*
Cyprus 16.03.1995
Cyprus 06.2008 (21)
Cyprus 16.03.1995
forma *friburgensis*
Switzerland, Jura 09.06.2021
Switzerland, Jura 09.06.2021

BELOW, ANTICLOCKWISE FROM BOTTOM LEFT
forma *friburgensis*
Switzerland, Jura 09.06.2021
Switzerland, Jura 09.06.2021
forma *trollii*
France, Aveyron 22.05.1996
England, Hampshire 18.06.1984 (20)

Ophrys schulzei Bornm. & Fleischm., *Mitth. Thüring. Bot. Vereins, n.f.,* 28: 60 (1911).

SYNONYMS. *Ophrys schulzei* var. *curdica* H.Fleischm. & Bornm.; *O. luristanica* Renz

DISTRIBUTION. SE Turkey to Lebanon and NW Iran.

HABITAT. Sparse grassland, scrub and oak woods in full sun to light shade on dry, calcareous soils; 500–1,700 m.

FLOWERING. April to June.

DISTINGUISHING FEATURES. 15–65 cm tall. Similar to *O. apifera* but with a shorter apex to the column, a much smaller lip in which the obtuse side lobes are larger relative to the mid-lobe, and a coherent speculum that covers most of the mid-lobe.

LEFT TO RIGHT
Turkey 09.05.1990 (8)
Turkey 13.05.1989 (8)
Lebanon 21.05.2002 (1)

Ophrys sphegodes Mill., *Gard. Dict. ed. 8*: no. 8 (1768). **Early spider orchid**

a. subsp. *sphegodes*

SYNONYMS. *Ophrys aranifera* Huds.; ?*O. exaltata* Ten.; *O. panormitana* (Tod.) Soó; *O. hebes* (Kalop.) E. & B.Willing; *O. delmeziana* P.Delforge; *O.* ×*poelmansiana* P.Delforge; *O. majellensis* (Helga Daiss & Herm.Daiss) P.Delforge; *O. cilentana* Devillers-Tersch. & Devillers; *O. classica* Devillers-Tersch. & Devillers; *O. tarquinia* P.Delforge; *O. negadensis* G. & W.Thiele; *O. zagoriana* G.Thiele & W.Thiele; *O. liburnica* Devillers & Devillers-Tersch.; *O. sphegodes* subsp. *grassoana* Cristaudo

DISTRIBUTION. W & S Europe from Spain, France and S England to Hungary and the Balkans.

HABITAT. Full sun to light shade in garigue, olive groves, chalk grassland and roadside verges; sea level to 1,300 m.

FLOWERING. March to June.

DISTINGUISHING FEATURES. 10–40 cm, or more, tall, slender. Sepals green (to white). Petals glabrous or with finely hairy margins. Lip medium brown (occasionally with a narrow yellow margin), flat to somewhat vaulted, entire to shallowly 3-lobed near the middle, 10–16 mm long, longer than the dorsal sepal, usually retuse around a small triangular to subulate, more or less porrect, point and with lateral margins that are fairly long-hairy towards the lip base; protuberances, if any, isolated from the lip margin and rarely more than ¼ as large as the stigmatic cavity; speculum usually H-shaped and connected to the lip base by distinct broad bands.

LEFT TO RIGHT
subsp. *sphegodes*
England, Sussex 12.05.2004 (25)
France, Aveyron 23.05.2008
Spain, Navarre 12.05.2014 (45)
Italy, Puglia 07.04.2016
Italy, Puglia, 14.04.2016
Italy, Puglia 18.05.2006

a.1. *Ophrys sphegodes* subsp. *sphegodes* var. *provincialis* (H. Baumann & Künkele) P.J. Cribb in Rolf Kühn *et al.*, *Field Guide Orchids Europe Medit.*: 241 (2019).

SYNONYM. *Ophrys sphegodes* subsp. *provincialis* H.Baumann & Künkele; *O. provincialis* (H.Baumann & Künkele) Paulus

DISTRIBUTION. SE France.

HABITAT. Full sun to light shade in olive groves, meadows, garigue, forest edges and open pine forest on calcareous soil; sea level to 700 m.

FLOWERING. April and May.

DISTINGUISHING FEATURES. 15–40 cm tall. Differs from the typical subsp. *sphegodes* in that the protuberances of the lip are absent or very weakly developed and in having a speculum that is always being delimited by a white line.

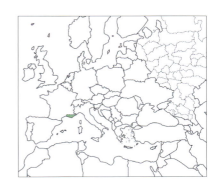

OPPOSITE PAGE, LEFT TO RIGHT
subsp. *sphegodes*
Italy, Puglia 13.04.2016
Italy, Puglia 11.04.2016
Italy, Tuscany 15.04.1999
Italy, Sicily 12.04.2022
Italy, Sicily 10.04.2022
Italy, Sicily 10.04.2022
France, Aveyron 20.05.2005
France, Aveyron 20.05.2005
France, Aveyron 20.05.2011

THIS PAGE, LEFT TO RIGHT
var. *provincialis*
France, Haute-Provence 14.05.2003
France, Provence 12.05.2007
France, Haute-Provence 14.05.2003
France, Haute-Provence 15.05.2006
France, Haute-Provence 11.05.2009

b. *Ophrys sphegodes* subsp. *aesculapii* (Renz)
Soó ex J.J.Wood, *Orchidee (Hamburg)* 31: 232 (1980).

SYNONYMS. *Ophrys aesculapii* Renz; *O. aranifera* subsp. *renzii* Soó; *O. renzii* Soó

DISTRIBUTION. S mainland Greece (including the Peloponnese) and Euboea.

HABITAT. Full sun to light shade in garigue and open maquis; sea level to 1,500 m.

FLOWERING. March to early May.

DISTINGUISHING FEATURES. 15–40 cm tall. Differs from the other 14 subspecies in having the following combination of features: lip (sub)entire, 9–14 mm long, dark brown with a 1.5–3 mm broad, yellow (to red-brown) margin, rounded to short-acuminate with a small terminal point and with lateral margins that are subglabrous to very short-hairy throughout; protuberances very low; speculum H-shaped; stigmatic cavity speckled green and brown, with the eye-like knobs pale (yellow-)green.

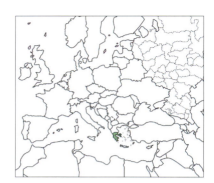

c. *Ophrys sphegodes* subsp. *amanensis*
(E.Nelson ex Renz & Taubenheim) H.A.Pedersen, P.J.Cribb & Rolf Kühn in Rolf Kühn *et al.*, *Field Guide Orchids Europe Medit.*: 242 (2019).

SYNONYMS. *Ophrys transhyrcana* subsp. *amanensis* E.Nelson ex Renz & Taubenheim; *O. amanensis* (E.Nelson ex Renz & Taubenheim) P.Delforge; *O. antalyensis* Kreutz & Seckel; *O. amanensis* subsp. *antalyensis* (Kreutz & Seckel) Kreutz

DISTRIBUTION. Endemic to CS Turkey.

HABITAT. In full sun to light shade in garigue, meadows and open woods of pine or oak on calcareous soil; sea level to 900 m.

FLOWERING. May and June.

DISTINGUISHING FEATURES. 20–60 cm tall. Differs from the other 14 subspecies in having the following combination of features: lateral sepals purplish-violet or bicoloured purplish-violet and pale violet, rarely white; lip large, 12–18 mm long, entire to distinctly 3-lobed with short-hairy lateral margins; protuberances as large as the stigmatic cavity.

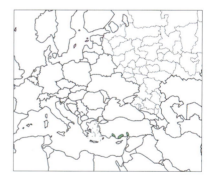

OPPOSITE PAGE, LEFT TO RIGHT
subsp. *aesculapii*
Greece, Peloponnese 27.04.2000 (2)
Greece, Attica 24.03.2018 (45)
(centre) Greece, Stirio-Kiriaki 09.04.2003 (3)
Greece, Peloponnese 20.04.2000 (56)
subsp. *amanensis*
Turkey, Antalya 13.05.2013 (40)
Turkey, Antalya 13.05.2013 (40)
Turkey, Antalya 13.05.2013 (40)
Turkey, Antalya 13.05.2013 (40)

d. *Ophrys sphegodes* subsp. *araneola* (Rchb.)

M. Laínz, *Anales Jard. Bot. Madrid* 40: 279 (1983).

SYNONYMS. *Ophrys araneola* Rchb.; *O. aranifera* subsp. *araneola* (Rchb.) K.Richt.; *O. litigiosa* E.G.Camus; *O. sphegodes* subsp. *litigiosa* (E.G.Camus) Bech.; *O. argentaria* Devillers-Tersch. & Devillers; ?*O. argensonensis* J.-C.Guérin & A.Merlet; *O. riojana* C.E.Hermos.; *O. sphegodes* var. *argentaria* (Devillers-Tersch. & Devillers) Faurh.; *O. illyrica* S. & K.Hertel; *O. ausonia* Devillers, Devillers-Tersch. & P.Delforge; *O. incantata* Devillers & Devillers-Tersch.; ?*O. maritima* Pacifico & Soca

DISTRIBUTION. SW & SC Europe from NE Spain to the Balkan peninsula and north to Germany.

HABITAT. Full sun to light shade in garigue, maquis, chalk grassland and roadside verges; sea level to 1,300 m.

FLOWERING. Mid-March to mid-April.

DISTINGUISHING FEATURES. A robust plant 10–45 cm tall. Similar to the typical subspecies from which it differs mainly in having a 6.5–10 mm long lip (vs. 10–16 mm in subsp. *sphegodes*) that is shorter than the dorsal sepal.

THIS PAGE, BELOW LEFT TO RIGHT
subsp. *araneola*
France, Aveyron 20.05.1999
Italy, Tuscany 30.04.2004
Italy, Tuscany 12.05.2002
Italy, Tuscany 13.05.2007

OPPOSITE PAGE, LEFT TO RIGHT
subsp. *araneola*
Switzerland, Baselland 22.04.1987
Switzerland, Baselland 22.04.1987
France, Aveyron 19.05.2003
France, Aveyron 22.05.1998
France, Aveyron 12.05.2006
France, Aveyron 14.05.2008
France, Aveyron 17.05.2009
France, Aveyron 14.05.2008
France, Aveyron 17.05.2006
France, Aveyron 18.05.2003

e. *Ophrys sphegodes* subsp. *atrata* (Rchb.f.)
A.Bolòs, *Veg. Com. Barcelona*: 265 (1950).
SYNONYMS. *Ophrys aranifera* var. *atrata* Rchb.f.;
O. incubacea Bianca; ?*O. brutia* P.Delforge;
O. incubacea var. *dianensis* Perazza & Doro;
O. incubacea subsp. *pacensis* F.M.Vázquez;
O. incubacea var. *septentrionalis* Perazza & R.Lorenz
DISTRIBUTION. S Europe from Portugal to W Balkan Peninsula.
HABITAT. Full sun to light shade in meadows, grassland, garigue, maquis and open forest; sea level to 1,300 m.
FLOWERING. March to May.
DISTINGUISHING FEATURES. 20–40 cm, or more, tall. Differs from the other 14 subspecies in having a lip with both well-developed, obliquely conical protuberances and a thick shaggy border of strikingly long hairs all around.

f. *Ophrys sphegodes* subsp. *aveyronensis*
J.J.Wood, *Orchidee (Hamburg)* 34: 106 (1983).
SYNONYMS. *Ophrys aveyronensis* (J.J.Wood) H.Baumann & Künkele; *O. vitorica* Kreutz
DISTRIBUTION. SC France and N Spain.
HABITAT. In full sun to light shade in grassland, hay meadows, scrubs and open oak groves, usually on calcareous soil; 400–900 m.
FLOWERING. May and June.
DISTINGUISHING FEATURES. 10–40 cm tall. Differs from the other 14 subspecies in having the following combination of features: sepals lilac-pink to dark rose-pink (rarely creamy-white); petals less than 2.5 times as long as broad, apricot or lilac-pink with green or reddish margins; lip at least as wide as long, entire to shallowly 3-lobed, with a usually marbled speculum and very low protuberances (if any).

OPPOSITE, LEFT TO RIGHT
subsp. *atrata*
France, Aveyron 20.05.1998
Italy, Puglia 11.04.2016
Spain, Majorca, Beliver 19.04.2013
Italy, Puglia 11.04.2016
Italy, Puglia 10.04.2016
Sardinia 09.04.2016
subsp. *aveyronensis*
France, Aveyron 25.05.2000
France, Aveyron 20.05.1989
France, Aveyron 15.06.2004

254　OPHRYS

THIS PAGE, LEFT TO RIGHT
subsp. *aveyronensis*
France, Aveyron 20.05.1998
France, Aveyron 20.05.1992
France, Aveyron 26.05.1994
France, Aveyron 14.05.2002
France, Aveyron 22.05.2000
France, Aveyron 25.05.2001

OPPOSITE PAGE, LEFT TO RIGHT
subsp. *aveyronensis*
France, Aveyron 20.05.2005
France, Aveyron 23.05.2006
France, Aveyron 18.05.1999
France, Lapanouse 20.05.1999
France, Aveyron 20.05.1987
France, Aveyron 19.05.2015
France, Aveyron 20.05.1997
France, Aveyron 15.05.1996
France, Aveyron 17.05.1998

g. *Ophrys sphegodes* subsp. *cretensis*
H.Baumann & Künkele, *Mitt. Arbeitskreis Heimische Orchid. Baden-Württemberg* 18: 375 (1986).

SYNONYMS. *Ophrys cretensis* (H.Baumann & Künkele) Paulus; *O. araneola* subsp. *cretensis* (H.Baumann & Künkele) Kreutz

DISTRIBUTION. Crete and Aegean Islands north to Chios.

HABITAT. Full sun to light shade in garigue, maquis and olive groves on calcareous, mostly dry soil; sea level to 1,000 m.

FLOWERING. February to mid-April.

DISTINGUISHING FEATURES. 20–50 cm tall. Differs from the other 14 subspecies in having the following combination of features: sepals pale green or bicoloured pale green and pale red-brown; lip entire, 5–9 mm long, rounded to short-acuminate with a small terminal point, rounded to truncate at base and with lateral margins that are subglabrous to short-hairy throughout; protuberances weakly developed to obliquely conical; speculum more or less H-shaped.

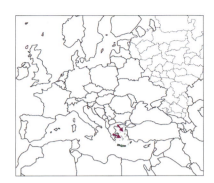

h. *Ophrys sphegodes* subsp. *epirotica* (Renz)
Gölz & H.R.Reinhard, *Mitt. Arbeitskreis Heimische Orchid. Baden-Württemberg* 15: 178 (1983).

SYNONYMS. *Ophrys aranifera* f. *epirotica* Renz; *O. epirotica* (Renz) Devillers-Tersch. & Devillers; *O. zeusii* M.Hirth; *O. taigetica* Presser & S.Hertel

DISTRIBUTION. Albania, Macedonia and Greece south to the Peloponnese.

HABITAT. Full sun to light shade in grassland, maquis and open forest; sea level to 1,500 m.

FLOWERING. Late April to mid-June.

DISTINGUISHING FEATURES. 15–40 cm tall. Similar to the earlier flowering subsp. *aesculapii* but differs in having a uniformly brown stigmatic cavity, with the eye-like knobs greyish blue.

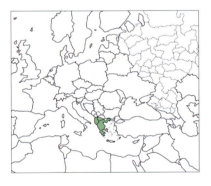

OPPOSITE, LEFT TO RIGHT FROM TOP
subsp. *cretensis*
Crete 09.04.2014
Crete 07.04.2015
Crete 05.04.2014
Crete 04.04.2016 (45)
subsp. *epirotica*
Greece, Arkadia 01.05.2000 (2)
Greece, Aroania 28.04.1999 (10)
Greece, Thessaly 06.05.1999 (2)

i. *Ophrys sphegodes* subsp. *gortynia*

H.Baumann & Künkele, *Mitt. Arbeitskreis Heimische Orchid. Baden-Württemberg* 18: 377 (1986).

SYNONYMS. *Ophrys gortynia* (H.Baumann & Künkele) Paulus; *O. mammosa* subsp. *gortynia* (H.Baumann & Künkele) H.Baumann & R.Lorenz

DISTRIBUTION. Crete, C Aegean Islands and possibly westernmost Anatolia.

HABITAT. Full sun to light shade in garigue, maquis and edges of forest on calcareous soil; sea level to 600 m.

FLOWERING. Mid-April and early May.

DISTINGUISHING FEATURES. 15–35 cm tall. Similar to subsp. *mammosa* but easily distinguished by its wedge-shaped (not rounded or truncate) lip base.

j. *Ophrys sphegodes* subsp. *helenae* (Renz) Soó & D.M.Moore, *Bot. J. Linn. Soc.* 76: 367 (1978).

SYNONYMS. *Ophrys helenae* Renz; *O. aranifera* subsp. *helenae* (Renz) Soó; *O. ×plakotiana* C., G., U. & W.Thiele

DISTRIBUTION. S Albania, S Macedonia and Greece south to the Peloponnese.

HABITAT. Full sun to light shade in grassland, maquis and open forest, mainly on calcareous soil; sea level to 1,000 m.

FLOWERING. April to June.

DISTINGUISHING FEATURES. 15–40 cm tall. A very distinctive orchid with a large, red-brown lip devoid of a speculum.

OPPOSITE, LEFT TO RIGHT FROM TOP
subsp. *gortynia*
Crete 03.04.2006 (45)
Crete 03.04.2006 (45)
Crete 03.04.2006 (45)
Crete 04.04.2015
Crete 04.04.2015
Crete 03.04.2006 (45)

BOTTOM ROW, LEFT TO RIGHT
subsp. *helenae*
Greece, Epirus 15.05.1971 (2)
Greece, Epirus 15.05.1971 (2)
Greece, Thessaly 06.05.1986 (11)
Greece, Viotia 29.04.1986 (11)

k. *Ophrys sphegodes* subsp. *lycia* (Renz & Taubenheim) H.A.Pedersen & P.J.Cribb in Rolf Kühn et al., *Field Guide Orchids Europe Medit.*: 255 (2019).
SYNONYM. *Ophrys lycia* Renz & Taubenheim.
DISTRIBUTION. Endemic to a small area in SW Turkey.
HABITAT. In full sun in garigue, grassland and open places in maquis on calcareous soil; 400–600 m.
FLOWERING. March and April.
DISTINGUISHING FEATURES. 15–40 cm, or more, tall. Differs from the other 14 subspecies in having the following combination of features: lateral sepals rose spotted purplish-violet or bicoloured rose and purplish violet; petals creamy-salmon; lip (oblong-)obtriangular with wedge-shaped base, a more or less H-shaped speculum and very low protuberances (if any).

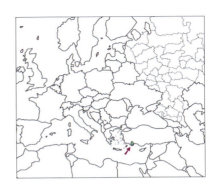

TOP ROW, LEFT TO RIGHT
subsp. *lycia*
Turkey, Antalya 14.04.2013 (47)
Turkey, Antalya 11.04.1982 (1)
Turkey, Antalya 04.04.1993 (8)

l. *Ophrys sphegodes* subsp. *passionis* (Sennen) Sanz & Nuet, *Guia Camp Orquíd. Catalunya*: 176 (1995).
SYNONYMS. *Ophrys passionis* Sennen; *O. sphegodes* subsp. *garganica* E.Nelson; *O. garganica* O. & E.Danesch; *O. garganica* subsp. *passionis* Paulus & Gack; *O. caloptera* Devillers-Tersch. & Devillers; *O. pseudoatrata* S.Hertel & Presser; ?*O. ligustica* Romolini & Soca; *O. minipassionis* Romolini & Soca
DISTRIBUTION. NE Spain across W & S France to Sardinia, Sicily and C & S mainland Italy.
HABITAT. Full sun to light shade in garigue, chalk grassland, olive groves and roadside verges; sea level to 1,300 m.
FLOWERING. March to June.
DISTINGUISHING FEATURES. 20–45 cm tall. Similar to the typical subspecies but differs in its often conspicuously broad petals, its dark brown lip colour and its speculum being basically H-shaped but with two additional short arms from the base.

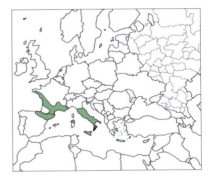

FROM SECOND ROW, LEFT TO RIGHT
subsp. *passionis*
Italy, Tuscany 18.04.1998
Italy, Tuscany 13.04.1999
Italy, Tuscany 18.04.2005
Italy, Tuscany 20.04.1999
Italy, Puglia 13.04.2016
Italy, Tuscany 18.04.1998

ABOVE, LEFT TO RIGHT
subsp. *passionis*
Italy, Tuscany 23.05.1996
Italy, Puglia 20.04.2010 (21)
Italy, Puglia 20.04.2008 (21)
Italy, Tuscany 18.04.1998
Italy, Tuscany 18.04.2005
Italy, Tuscany 18.04.1998

m. *Ophrys sphegodes* subsp. *sipontensis*
(Kreutz) H.A.Pedersen & Faurh., *J. Eur. Orch.* 37: 289 (2005).

SYNONYMS. *O. garganica* subsp. *sipontensis* Kreutz; ?*O. murgiana* Cillo, Medagli & Margh.

DISTRIBUTION. Endemic to S mainland Italy (Puglia and Campania).

HABITAT. Full sun in poor, dry to humid grassland and garigue on calcareous soil, often associated with asphodels; sea level to 900 m.

FLOWERING. March to early May.

DISTINGUISHING FEATURES. 15–60 cm tall. Reminiscent of subsp. *amanensis*, but distinguished by its never bicoloured lateral sepals and its consistently entire lip that has long-hairy lateral margins and only very weakly developed protuberances, if any.

BELOW, LEFT TO RIGHT
subsp. *sipontensis*
Italy, Puglia 14.03.2009 (21)
Italy, Puglia 04.2020 (21)
Italy, Puglia 23.04.2008 (21)
Italy, Puglia 10.04.2016
Italy, Puglia 04.2004 (21)

n. *Ophrys sphegodes* subsp. *spruneri* (Nyman)
E.Nelson, *Gestaltw. Artb. Orchid. Eur. Mittelmeerl.*: 183 (1962).

SYNONYMS. *Ophrys spruneri* Nyman; *O. ferrum-equinum* subsp. *spruneri* (Nyman) E.G.Camus; *O. grigoriana* G. & H.Kretzschmar; *O. spruneri* subsp. *grigoriana* (G. & H.Kretzschmar) H.Kretzschmar

DISTRIBUTION. Ionian Islands and W mainland Greece across the Peloponnese to Crete and a few Aegean Islands.

HABITAT. Full sun to light shade in garigue, maquis and open forest on calcareous soil; sea level to 1,100 m.

FLOWERING. Late February to early May.

DISTINGUISHING FEATURES. 10–50 cm tall. Reminiscent of subsp. *amanensis*, but distinguished by its almost consistently 3-lobed lip that has only weakly developed protuberances, if any.

BELOW, LEFT TO RIGHT
subsp. *spruneri*
Greece 06.04.2015
Greece 06.04.2015
Crete 06.04.2015
Crete 02.04.2014
Crete 02.02.2014
Greece, Peloponnese 20.03.2015 (21)

o. ***Ophrys sphegodes* subsp. *taurica*** (Aggeenko) Soó ex Niketic & Djordjevic, *Bull. Nat. Hist. Mus. Belgrade* 11: 106 (2018).

SYNONYMS. *Ophrys mammosa* Desf.; *O. aranifera* var. *taurica* Aggenko; *O. sintenisii* H.Fleischm. & Bornm.; *O. aranifera* subsp. *macedonica* H.Fleischm. ex Soó; *O. pseudomammosa* Renz; *O. caucasica* Woronow; *O. taurica* (Aggeenko) Nevski; *O. sphegodes* subsp. *mammosa* (Desf.) Soó ex E.Nelson; *O. sphegodes* subsp. *sintenisii* (H.Fleischm. & Bornm.) E.Nelson; *O. grammica* (B. & E.Willing) Devillers-Tersch. & Devillers; *O. caucasica* subsp. *cyclocheila* Aver.; *O. macedonica* (H.Fleischm. ex Soó) Devillers-Tersch. & Devillers; *O. cyclocheila* (Aver.) P.Delforge; *O. leucophthalma* Devillers-Tersch. & Devillers; *O. hittitica* Kreutz & Ruedi Peter; *O. iceliensis* Kreutz; *O. alasiatica* Kreutz, Segers & H.Walraven; *O. janrenzii* M.Hirth; *O. amanensis* subsp. *iceliensis* (Kreutz) Kreutz; *O. morio* Paulus & Kreutz; *O. transhyrcana* subsp. *morio* (Paulus & Kreutz) Kreutz; *O. transhyrcana* subsp. *sintenisii* (H.Fleischm. & Bornm.) Kreutz; *O. hystera* Kreutz & Ruedi Peter; *O. mammosa* subsp. *mouterdeana* B. & H.Baumann; *O. mammosa* subsp. *posteria* B. & H.Baumann; *O. hansreinhardii* M.Hirth; *O. sphegodes* subsp. *catalcana* Kreutz; *O. tremoris* Gämperle & Gölz; *O. knossia* (A.Alibertis) P.Delforge

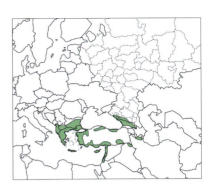

BELOW, LEFT TO RIGHT
subsp. *taurica*
Crete 29.03.2014
Cyprus 15.03.2008
Cyprus 12.03.2002

DISTRIBUTION. From S Balkan Peninsula across Turkey to the Caucasus, S Turkmenistan, Cyprus and Israel.

HABITAT. Full sun to light shade in garigue, maquis, grassland, roadside verges, olive groves and open pine forest; sea level to 1,400 m.

FLOWERING. March to June.

DISTINGUISHING FEATURES. 15–70 cm tall. Differs from the other 14 subspecies in having the following combination of features: lateral sepals usually bicoloured green and red-brown; lip entire to shallowly 3-lobed, 9–20 mm long, rounded to short-acuminate with a small terminal point, rounded to truncate at base and with lateral margins that are subglabrous to short-hairy throughout; protuberances (usually) conspicuous, obliquely conical and at least half as large as the stigmatic cavity; speculum H-shaped or consisting of two parallel bands.

NOTE. This subspecies is extremely variable, as reflected by the comprehensive synonymy.

LEFT TO RIGHT FROM TOP
subsp. *taurica*
Cyprus 14.03.2003
Cyprus, Troodos 19.03.2002
Crete 05.04.2014
Cyprus 18.03.2009
Cyprus 15.03.2010
Cyprus 15.03.2010
Cyprus 10.03.2008
Cyprus 14.03.2009
Cyprus, Akamas 15.03.1998
Cyprus 14.03.2000
Cyprus 15.03.1997
Cyprus, Troodos 18.03.2011
Cyprus 24.03.2001
Cyprus 18.03.2009
Cyprus 18.03.2004
Cyprus 20.03.2007
Cyprus 18.03.2004

LEFT TO RIGHT FROM TOP
subsp. *taurica*
Cyprus 16.03.2002
Cyprus 16.03.2002
Cyprus 21.03.2007
Cyprus 18.03.1999
Cyprus, north 14.04.2011
Cyprus 21.03.2001
Crete 04.04.2015
Cyprus 12.03.1999

o.1. *Ophrys sphegodes* subsp. *taurica* var. *transhyrcana* (Czerniak.) P.J. Cribb in Rolf Kühn et al., Orchids Europe Medit.: 259 (2019).

SYNONYMS. *Ophrys transhyrcana* Czerniak; *Ophrys mammosa* Desf. subsp. *transhyrcana* (Czerniak.) Buttler: *O. sphegodes* Mill. subsp. *transhyrcana* (Czerniak.) Soó

Mainly characterised by a conspicuously drawn-out column apex. Distributed in the E Mediterranean region.

BELOW, LEFT TO RIGHT
subsp. *taurica* var. *transhyrcana*
Cyprus, Aghios Nikolaos 14.03.2011
Cyprus 14.03.2003
Cyprus, Pano Polemedia 18.03.2001
Cyprus, Aghios Nikolaos 18.03.2011
Cyprus, Akamas 17.03.1999
Cyprus, Aghios Nikolaos 20.03.2011

Ophrys bertolonii Moretti, *Quibus. Pl. Ital.* 6: 9 (1823).

SYNONYMS. *Arachnites bertolonii* (Moretti) Tod.; *O. dalmatica* (Murr) Soó; *O. marzensis* Soca; *O. romolinii* Soca

DISTRIBUTION. Sicily, mainland Italy, Malta and former Jugoslavia.

HABITAT. In dry to moist, calcareous grassland, garigue, olive groves, roadside verges, and open woods; sea level to 1,200 m.

FLOWERING. March to June.

DISTINGUISHING FEATURES. 10–35 cm tall. Sepals rose-pink to greenish white. Differs from other species with an acute column by its distinctive saddle-shaped lip with a shiny bluish speculum in the apical half and by its column tapering towards base (in side view).

CLOCKWISE FROM LEFT
Italy, Puglia 15.04.2016
Sicily 11.04.2018
Sicily 12.04.2018
Sicily 12.04.2018
Italy, Tuscany 14.04.1997

Ophrys lunulata Parl., *Giorn. Sci. Sicilia* 62: 4 (1838).

SYNONYMS. *Arachnites lunulatus* (Parl.) Tod.; *Ophrys sphegodes* subsp. *lunulata* (Parl.) H.Sund.

DISTRIBUTION. Malta (extinct?), Sicily.

HABITAT. In grassland, garigue and open woodland on dry to fairly moist, calcareous to slightly acidic soils in full sun to light shade; sea level to 800 m.

FLOWERING. March and April.

DISTINGUISHING FEATURES. 10–40 cm tall. Characterised by its long, narrow petals and deeply 3-lobed, apparently narrow lip (actually with strongly reflexed margins) with a crescent-shaped or sometimes fragmented, centrally placed speculum.

LEFT TO RIGHT
Sicily 10.04.2018
Sicily 10.04.2018
Sicily 10.04.2018

Ophrys ferrum-equinum Desf., *Ann. Mus. Hist. Nat.* 10: 226 (1807). **Horseshoe orchid**

a. subsp. *ferrum-equinum*

SYNONYM. *Ophrys labiosa* Kreutz

DISTRIBUTION. S Balkans across the Greek islands (absent from Crete) to SW & S Turkey.

HABITAT. On calcareous, dry to moist soil in full sun to light shade, in grassland, garigue, open pine woodland and olive groves; sea level to 1,100 m.

FLOWERING. March to May.

DISTINGUISHING FEATURES. 10–35 cm tall. Sepals and petals pink to rose-purple, rarely white; petals subglabrous. Lip entire to moderately 3-lobed, straight, with black or very dark brown ground colour and velvety hairs in its marginal part, in its distal part, with spreading to slightly recurved sides (making it appear elliptic to obovate in outline, widest above the middle), ending in a small, triangular, more or less porrect point; protuberances, if any, very low and isolated from the lip margin; speculum isolated from lip base, consisting of a horseshoe- to butterfly-shaped mark or two parallel bands. Superficially similar to *O. sphegodes* subsp. *spruneri*, which differs in having an H-shaped speculum with distinct connections to the base of the lip.

NOTE. In a few Aegean isles, plants differing from typical *O. ferrum-equinum* subsp. *ferrum-equinum* in lip and speculum shape are recognised by certain authors as *O. icariensis* M.Hirth & H.Spaeth, whereas other deviant Greek populations are sometimes recognised as *O. olympiotissa* Paulus. We think that all such plants are hybrids, with *O. ferrum-equinum* subsp. *ferrum-equinum* constituting one of the parental taxa.

b. *O. ferrum-equinum* subsp. *gottfriediana*

(Renz) E.Nelson, *Gestaltw. Artb. Orchid. Eur. Mittelmeerl.*: 141 (1962).

SYNONYMS. *Ophrys gottfriediana* Renz; *O. spruneri* subsp. *gottfriediana* (Renz) Soó; *O. ferrum-equinum* var. *pseudogottfriediana* Paulus

DISTRIBUTION. Endemic to Greece where it mainly occurs in the Ionian Islands.

HABITAT. As for the species; sea level to 700 m.

FLOWERING. March to May.

DISTINGUISHING FEATURES. Distinguished from the typical subspecies in that the sepals and petals are often muddy rose-coloured, olive-green or white and in that the lip has reflexed sides in its distal part (making it appear triangular in outline, widest below the middle).

OPPOSITE, LEFT TO RIGHT

subsp. *ferrum-equinum*
Greece, Corfu 03.1981 (22)
Greece, Peloponnese 30.03.1991 (1)
Greece, Peloponnese 28.03.2015 (21)
Greece, Peloponnese 23.03.2015 (21)
Greece, Peloponnese 24.03.2015 (21)
Greece, Peloponnese 22.03.2015 (21)
Greece, Lesbos 29.03.2018 (45)

subsp. *gottfriediana*
Greece, Corfu 05.1992 (22)
Greece, Corfu 15.04.2001 (1)

Ophrys argolica H.Fleischm., *Verh. K. K. Zool.-Bot. Ges. Wien* 69: 295 (1919). **Eyed bee orchid**

SYNONYM. *Ophrys ferrum-equinum* subsp. *argolica* (H.Fleischm.) Soó

a. subsp. *argolica*

DISTRIBUTION. Endemic to Greece where it mainly occurs in the Peloponnese and nearby areas.

HABITAT. Full sun to light shade in garigue, grassland, olive groves and open forest on calcareous soil; sea level to 1,300 m.

FLOWERING. March to May.

DISTINGUISHING FEATURES. 10–30 cm, or more, tall. Sepals spreading, rose to pink. Petals hairy, less than 3 times as long as broad. Lip entire to shallowly 3-lobed, straight, widest above the middle, slightly vaulted, with (red-)brown ground colour and conspicuous white hairs in its marginal part (especially towards the base), ending in a 1–2 mm broad, more or less porrect, triangular point; protuberances, if any, very low and isolated from the lip margin; speculum isolated from the lip base or only connected to it by delicate lines, consisting of a butterfly- to horseshoe-shaped mark or a pair of spots. Stigmatic cavity 2–4 mm wide, brownish with pale centre and often a dark brown line across.

TOP TWO ROWS, CLOCKWISE FROM TOP LEFT
subsp. *argolica*
Greece, Peloponnese 04.1995 (25)
Greece, Peloponnese 27.03.2015 (21)
Greece, Peloponnese 04.1995 (25)
Greece, Peloponnese 04.1989 (22)
Greece, Peloponnese 04.1995 (22)
Greece, Peloponnese 30.03.1989 (20)
Greece, Peloponnese 30.03.1989 (20)

b. *Ophrys argolica* subsp. *aegaea* (Kalteisen & H.R.Reinhard) H.A.Pedersen & Faurh., *Orchidee (Hamburg)* 53: 345 (2002).

SYNONYMS. *Ophrys aegaea* Kalteisen & H.R.Reinhard; *O. ferrum-equinum* subsp. *aegaea* (Kalteisen & H.R.Reinhard) H.Baumann & R.Lorenz

DISTRIBUTION. Endemic to Greece where it seems to be restricted to Karpathos and a few Cycladean islands.

HABITAT. Full sun to light shade in pine woodland, olive groves and asphodel-dominated garigue; sea level to 700 m.

FLOWERING. February to April.

DISTINGUISHING FEATURES. 10–20 cm, or more, tall. Similar to the typical subspecies but differs in having a more compact habit and a lip that is marginally grey- to light brown-hairy (not white-hairy) in the basal part and that ends in a 0.5–1 mm broad terminal point.

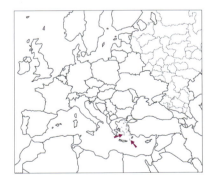

BOTTOM ROW, ALL IMAGES
subsp. *aegaea*
Greece, Karpathos 20.03.2000 (50)

c. *Ophrys argolica* subsp. *biscutella* (O.Danesch & E.Danesch) Kreutz, *Kompend. Eur. Orchid.*: 85 (2004).
SYNONYMS. *Ophrys biscutella* O. & E.Danesch; *O. exaltata* subsp. *sundermannii* Soó, nom. inval.
DISTRIBUTION. S mainland Italy, Croatia (Korcula).
HABITAT. Full sun to light shade in grassland, garigue, maquis and open forest on calcareous soil; sea level to 1,300 m.
FLOWERING. April to May.
DISTINGUISHING FEATURES. 10–60 cm tall. Differs from the other seven subspecies in having the following combination of features: sepals reflexed, white to red-purple (rarely green); petals up to 3 times as long as wide; lip entire, broadest above the middle, with spreading sides, (red-)brown in the central part, ending in a 1–3 mm broad, triangular point; lateral margins whitish- to light brown-hairy towards lip base; stigmatic cavity 3–4 mm wide, brown.

d. *Ophrys argolica* subsp. *climacis* (Heimeier & Perschke) H.A.Pedersen & P.J.Cribb in Rolf Kühn et al., *Field Guide Orchids Europe Medit.*: 270 (2019).
SYNONYMS. *Ophrys climacis* Heimeier & Perschke; *O. ferrum-equinum* subsp. *climacis* (Heimeier & Perschke) H.Baumann & R.Lorenz
DISTRIBUTION. SW Turkey (Antalya).
HABITAT. In light shade in open pine and cypress forest (often on virtually bare ground) and in the edges of maquis, always on calcareous soil; sea level to 1,200 m.
FLOWERING. Mainly in March and April.
DISTINGUISHING FEATURES. 15–40 cm tall. Differs from the other seven subspecies in having the following combination of features: lateral sepals narrowly and obliquely ovate, at least twice as long as wide, (whitish-)green, occasionally suffused with rose; petals obliquely lanceolate, more than three times as long as wide, green to yellow; lip entire or 3-lobed, usually with a broad yellow margin in its distal part.

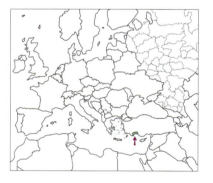

LEFT TO RIGHT FROM TOP
subsp. *biscutella*
Italy, Puglia 09.04.2016
Italy, Puglia 09.04.2016
Italy, Tuscany 15.04.2012
Italy, Puglia 11.04.2016
Italy, Puglia 12.04.2016
Italy, Tuscany 12.04.2005
subsp. *climacis*
Turkey, Antalya 08.03.1998 (56)
Turkey, Antalya 08.03.1998 (56)
Turkey, Antalya 08.03.1998 (56)

e. *Ophrys argolica* subsp. *crabronifera* (Sebast. & Mauri) Faurh., *Orchidee (Hamburg)* 53: 345 (2002).

SYNONYMS. *Ophrys crabronifera* Sebast. & Mauri; *O. fuciflora* subsp. *pollinensis* E.Nelson, nom. inval.; *O. pollinensis* Devillers-Tersch. & Devillers; *O. argolica* subsp. *pollinensis* (E.Nelson) Kreutz

DISTRIBUTION. Endemic to SC Italy.

HABITAT. Full sun to shade in grassland, garigue, maquis and woodland on calcareous soil; sea level to 1,000 m.

FLOWERING. March to May.

DISTINGUISHING FEATURES. 20–65 cm tall. Similar to subsp. *biscutella* but distinguished by distinctly recurved sides of the lip and a more than 4 mm wide stigmatic cavity.

BELOW, LEFT TO RIGHT
subsp. *crabronifera*
Italy, Tuscany 08.04.2003
Italy, Tuscany 09.04.2006
Italy, Umbria 22.04.2008 (21)
Italy, Umbria 14.04.2008 (21)
Italy, Tuscany 13.04.2006
Italy, Tuscany 09.04.2006
Italy, Tuscany 13.04.1999
Italy, Tuscany 09.04.2006

f. *Ophrys argolica* subsp. *elegans* (Renz) E.Nelson, *Gestaltw. Artb. Orchid. Eur. Mittelmeerl.*: 153 (1962).

SYNONYMS. *Ophrys gottfriediana* subsp. *elegans* Renz; *O. elegans* (Renz) H.Baumann & Künkele

DISTRIBUTION. Cyprus.

HABITAT. Full sun to light shade in garigue, maquis and open pine forest; sea level to 1,400 m.

FLOWERING. February to April.

DISTINGUISHING FEATURES. 5–20 cm tall. Easily distinguished from the other seven subspecies in having the combination of reflexed sepals and a 3-lobed, strongly vaulted lip.

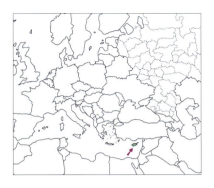

BELOW, TOP TO BOTTOM, LEFT TO RIGHT
subsp. *elegans*
Cyprus 08.03.1999
Cyprus 13.03.1999
Cyprus 15.03.1999
Cyprus 13.03.2001
Cyprus 15.03.1999
Cyprus 19.03.2009
Cyprus 16.03.2000
Cyprus 14.03.2011
Cyprus 11.03.1997

g. *Ophrys argolica* subsp. *lesbis* (Gölz & H.R.Reinhard) H.A.Pedersen & Faurh., *Orchidee (Hamburg)* 53: 345 (2002).

SYNONYMS. *Ophrys lesbis* Gölz & H.R.Reinhard; *O. ferrum-equinum* subsp. *lesbis* (Gölz & H.R.Reinhard) H.Baumann & R.Lorenz

DISTRIBUTION. NW Lesbos to SW Turkey.

HABITAT. Full sun to light shade in garigue, maquis and open oak forest on calcareous soil; sea level to 300 m.

FLOWERING. Mid-March to late-April.

DISTINGUISHING FEATURES. 8–20 cm, or more, tall. Similar to subsp. *lucis* but differs in having an entire lip that is broadest above the middle.

h. *Ophrys argolica* subsp. *lucis* (Kalteisen & H.R.Reinhard) H.A.Pedersen & Faurh., *Orchidee (Hamburg)* 53: 345 (2002).

SYNONYMS. *Ophrys aegaea* subsp. *lucis* Kalteisen & H.R.Reinhard; *O. lucis* (Kalteisen & H.R.Reinhard) Paulus & Gack; *O. ferrum-equinum* subsp. *lucis* (Kalteisen & H.R.Reinhard) H.Baumann & R.Lorenz; *O. argolica* subsp. *atargatis* Kreutz

DISTRIBUTION. SE Aegean Islands to S Turkey and N Syria.

HABITAT. Full sun to light shade in garigue and open pine forest on calcareous soil; sea level to 1,000 m.

FLOWERING. February to mid-April.

DISTINGUISHING FEATURES. 10–30 cm tall. Differs from the other seven subspecies in having the following combination of features: sepals spreading, white to red-purple (rarely green); petals less than 3 times as long as broad; lip 3-lobed, broadest at or below the middle, slightly vaulted, olive-green to red-brown in the central part, ending in a 1–2 mm broad, triangular point; lateral margins grey- to brown-hairy towards the lip base; stigmatic cavity 3–4 mm wide, brownish with a dark brown line across.

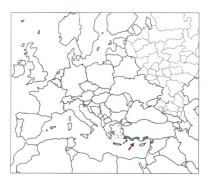

OPPOSITE, LEFT TO RIGHT FROM TOP
subsp. *lesbis*
Greece, Lesbos 26.03.2018 (45)
Greece, Lesbos 26.03.2018 (45)
Greece, Lesbos 09.04.1998 (1)
Greece, Lesbos 26.03.2018 (45)
Greece, Lesbos 26.03.2018 (45)
subsp. *lucis*
Greece, Rhodes 07.04.1995 (1)
Greece, Rhodes 15.03.2000 (50)
Greece, Rhodes 15.04.1982 (8)
Greece, Rhodes 15.03.2000 (50)

Ophrys holosericea (Burm.f.) Greuter, *Boissiera* 13: 185 (1967). **Late spider orchid**

SYNONYMS. *Orchis holosericea* Burm.f.

NOTE. We use here the name *Ophrys holosericea* rather than *Ophrys fuciflora* which appeared in the first edition. In doing so, we follow Greuter (2008) who argued for the acceptance of the former over the latter and also over *Ophrys holoserica*. He considered that *O. fuciflora* Haller of 1768 was not validly published while *O. holoserica* was an orthographic error.

a. subsp. *holosericea*

SYNONYMS. *Orchis fuciflora* Crantz; *Arachnites fuciflorus* F.W.Schmidt; *Ophrys fuciflora* (F.W.Schmidt) Moench; *O. episcopalis* Poir.; *O. fuciflora* (Crantz) Rchb.f., nom. illeg.; *O. annae* Devillers-Tersch. & Devillers; *O. aegirtica* P.Delforge; *O. aramaeorum* P.Delforge; *O. serotina* Rolli ex Paulus; *O. helios* Kreutz; *O. lyciensis* Paulus, D. & U. Rückbr.; *O. halia* Paulus; *O. lacaena* P.Delforge; *O. medea* Devillers & Devillers-Tersch.; *O. colossaea* P.Delforge; *O. druentica* P.Delforge & Viglione; *O. chiosica* P.Delforge; *O. holosericea* subsp. *cyrenaica* Kreutz; *O. holosericea* subsp. *taloniensis* Kreutz; *O. saliarisii* Paulus & M.Hirth; *O. holosericea* subsp. *vanbruggeniana* J. & L.Essink & Kreutz; *O. appennina* Romolini & Soca; *O. cinnabarina* Romolini & Soca; *O. pinguis* Romolini & Soca

DISTRIBUTION. From SE England and France across SC and S Europe to Turkey, Israel and NE Libya; absent from the Iberian peninsula, the Balearic Islands and Cyprus.

HABITAT. Dry to moist, often calcareous soils, in full sun to light shade, in meadows, garigue, roadsides, open oak and pine woodland and olive groves; sea level to 1,500 m.

FLOWERING. Late March to July, earlier in the Mediterranean, later in the north.

DISTINGUISHING FEATURES. 10–65 cm (or more) tall, bearing flowers along more than half of the stem's length. Leaves still fresh at the peak of flowering. Sepals rose to red-purple (rarely white or green), 9–16 mm long, with flat margins, up to 3 times as long as the petals. Petals (3–)4–6 mm long, distinctly longer than broad, with spreading hairs. Lip vertically oriented or nearly so, with spreading to recurved sides, entire or occasionally shallowly 3-lobed, ending

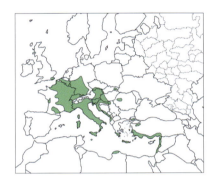

OPPOSITE, LEFT TO RIGHT FROM TOP
subsp. *holosericea*
Germany, Baden-Württemberg 09.05.2020
Germany, Baden-Württemberg 09.05.2020
France, Alsace 14.05.2020
France, Alsace 20.05.2020
France, Alsace 20.05.2020
France, Alsace 20.05.2020
Crete 10.04.2014
France, Haute Provence 16.05.1995
France, Provence 14.05.2012
France, Alsace 31.05.2016
France, Alsace 14.05.2021

in an upcurved, broad and conspicuous appendage that is rectangular, rhombic or obtriangular and often dentate; protuberances weakly developed to obliquely conical, distinctly isolated from the lip margin; speculum at least two-thirds as broad as the lip, consisting of coherent, well-defined, usually complicated marks, connected to the lip base, not delimited by a broad, unbranched, cream band; protuberances bicoloured brown and cream to brown and pale green. Column acute.

NOTE. In mainland Italy, from S Lazio to N Calabria, some populations flower relatively late and consist of generally tall plants with small flowers in a lax inflorescence. They may be referred to as var. *gracilis* (Büel, O.Danesch & E.Danesch) Faurh., although this variety does not seem to be clearly delimited from the typical form.

b. *Ophrys holosericea* subsp. *andria* (P.Delforge) Faurh., *Orchidee* (Hamburg) 53: 345 (2002).

SYNONYMS. *Ophrys andria* P.Delforge; *O. aeoli* P.Delforge; *O. thesei* P.Delforge; *O. fuciflora* subsp. *andria* (P.Delforge) Faurh.

DISTRIBUTION. Endemic to Greece where it seems to be restricted to the Cyclades.

HABITAT. Full sun in garigue on calcareous soil; sea level to 400 m.

FLOWERING. Mid-March to mid-April.

DISTINGUISHING FEATURES. 15–50 cm tall. Similar to the typical subspecies but differs in having a strongly reduced and fragmented speculum consisting of isolated spots and dashes (rarely more complex but still isolated marks).

OPPOSITE, LEFT TO RIGHT FROM TOP
subsp. *holosericea*
France, Alsace 07.06.2016
France, Alsace 16.06.2016
France, Alsace 26.05.2017
France, Alsace 13.06.2016
France, Alsace 09.06.2021
France, Alsace 23.05.2017
France, Alsace 31.05.2016
France, Alsace 23.05.2017
France, Alsace 09.05.2021
France, Alsace 14.05.2021
France, Alsace 28.05.2021

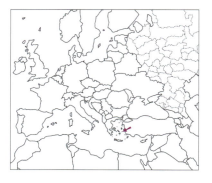

BELOW, LEFT TO RIGHT
subsp. *andria*
Greece, Siros 14.03.2003 (1)
Greece, Kapaia Andros 06.04.2005 (56)
Greece, Kapaia Andros 06.04.2005 (56)

c. ***Ophrys holosericea* subsp. *apulica*** (O.Danesch & E.Danesch) Buttler, *Willdenowia* 16: 115 (1986).

SYNONYMS. *Ophrys fuciflora* subsp. *apulica* O.Danesch & E.Danesch; *O. apulica* (O.Danesch & E.Danesch) O.Danesch & E.Danesch;?*O. dinarica* Kranjcev & P.Delforge; ?*O. pharia* Devillers & Devillers-Tersch.

DISTRIBUTION. Sicily and S to SE mainland Italy; possibly also in S Croatia.

HABITAT. Full sun to light shade in grassland, garigue, maquis and open forest on calcareous soil; sea level to 900 m.

FLOWERING. Late March to early June.

DISTINGUISHING FEATURES. 15–35 cm, or more, tall. Similar to large-flowered forms of the typical subspecies but differs in having (4–)6–9 mm long petals and a lip with sigmoidly curved sides and a straight or upcurved terminal appendage.

LEFT TO RIGHT, FROM TOP LEFT
subsp. *apulica*
Italy, Puglia 10.04.2016
Italy, Puglia 10.04.2016
Italy, Puglia 12.04.2016
Italy, Puglia 11.04.2016
Italy, Puglia 11.04.2016
Italy, Puglia 13.04.2016
Italy, Puglia 11.04.2016
Italy, Puglia 11.04.2016
Italy, Puglia 13.04.2016
Italy, Puglia 07.04.2016
Italy, Puglia 07.04.2016
Italy, Puglia 11.04.2016

d. *Ophrys holosericea* subsp. *biancae* (Tod.)

Faurh. & H.A.Pedersen, *Orchidee* (Hamburg) 53: 345 (2002).

SYNONYMS. *Arachnites biancae* Tod.; *Ophrys biancae* (Tod.) Macch.; ?*O.* ×*anapei* O. & E.Danesch; *O. oxyrrhynchos* subsp. *biancae* (Tod.) Galesi, Cristaudo & Maugeri; *O. fuciflora* subsp. *biancae* (Tod.) Faurh. & H.A.Pedersen.

DISTRIBUTION. Sicily.

HABITAT. Full sun to light shade in grassland, garigue, maquis and open forest on calcareous soil; sea level to 700 m.

FLOWERING. March and April.

DISTINGUISHING FEATURES. 10–25 cm tall. Similar to subsp. *oxyrrhynchos* but differs in having a trapeziform (not flabellate) lip, the central part of which is red- to yellow-brown (not dark brown).

e. *Ophrys holosericea* subsp. *bornmuelleri* (M.Schulze) H. Sund., *Taxon* 24: 625 (1975).

SYNONYMS. *Ophrys bornmuelleri* M.Schulze; *O. fuciflora* subsp. *bornmuelleri* (M.Schulze) B.Willing & E.Willing; *O. carduchorum* (Renz & Taubenheim) P.Delforge; *O. ziyaretiana* Kreutz & Ruedi Peter; *O. bornmuelleri* subsp. *ziyaretiana* (Kreutz & Ruedi Peter) Kreutz; *O. fuciflora* var. *ziyaretiana* (Kreutz & Ruedi Peter) Faurh. & H.A.Pedersen; *O. aphrodite* Devillers & Devillers-Tersch.

DISTRIBUTION. Cyprus, CS Turkey to Israel and N Iraq.

HABITAT. Light shade in maquis, olive groves and open forest, less often in full sun in garigue or on roadside verges; sea level to 1,200 m.

FLOWERING. Mid-March to mid-April.

DISTINGUISHING FEATURES. 15–50 cm tall. Differs from the other 14 subspecies in having the following combination of features: sepals green to white (rarely rose); petals 1.5–3(–4) mm long, about as long as broad; lip usually closer to horizontal than vertical orientation, trapeziform to square; terminal appendage less than half as long as the column; speculum well-developed, connected to the lip base by distinct bands.

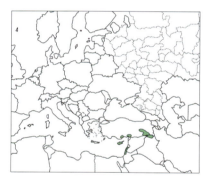

LEFT TO RIGHT, FROM TOP
subsp. *biancae*
Sicily 21.03.2003 (2)
Sicily 21.03.2003 (2)
Sicily 21.03.2003 (2)
subsp. *bornmuelleri*
Cyprus 18.03.2002
Cyprus 12.03.2009
Cyprus 26.03.1994
Cyprus 18.03.2004
Cyprus 13.03.2005
Cyprus 20.03.1996

f. *Ophrys holosericea* subsp. *candica* (E.Nelson ex Soó) Renz & Taubenheim, *Notes Roy. Bot. Gard. Edinburgh* 41: 276 (1983).

SYNONYMS. *Ophrys fuciflora* subsp. *candica* E.Nelson ex Soó; *O. candica* (E.Nelson ex Soó) H.Baumann & Künkele; *O. minoa* (C.Alibertis & A.Alibertis) P.Delforge; ?*O. calliantha* Bartolo & Pulv.; *O. cytherea* (B.Baumann & H.Baumann) P.Delforge

DISTRIBUTION. From Sicily and S mainland Italy across S Greece to SW Turkey.

HABITAT. Full sun to light shade in garigue and maquis, usually on calcareous soil; sea level to 1,200 m.

FLOWERING. April and May.

DISTINGUISHING FEATURES. 15–45 cm tall. Similar to the typical subspecies but differs in having a pronouncedly marbled speculum that is delimited by a broad, virtually unbranched, cream band.

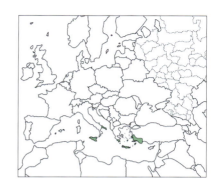

g. *Ophrys holosericea* subsp. *chestermanii* J.J.Wood, *Orchidee* (Hamburg) 33: 67 (1982).

SYNONYMS. *Ophrys chestermanii* (J.J.Wood) Gölz & H.R.Reinhard; *O. fuciflora* subsp. *chestermanii* (J.J.Wood) H.Blatt & W.Wirth.

DISTRIBUTION. Sardinia.

HABITAT. Light shade on moist, calcareous soil; typically growing on mossy slopes under holm oak (*Quercus ilex*); sea level to 600 m.

FLOWERING. Late March to mid-May.

DISTINGUISHING FEATURES. 10–30 cm tall. Similar to the typical subspecies but differs in having a dark brown to purple-black, usually trapeziform lip with a small, simple, usually H-shaped speculum that is less than half as wide as the lip.

LEFT TO RIGHT, FROM TOP
subsp. *candica*
Rhodes 11.04.1998 (50)
Rhodes 03.04.2001 (2)
Crete 17.04.2011 (21)
Rhodes 11.04.1998 (50)

subsp. *chestermanii*
Sardinia 20.04.1997 (45)
Sardinia 20.04.1997 (45)
Sardinia 20.04.1997 (45)
Sardinia 05.05.1994 (1)
Sardinia 20.04.1997 (45)

h. *Ophrys holosericea* subsp. *elatior* (Paulus)
H.Baumann & Künkele, *Farn-Blütenpfl. Baden-Würt.* 8: 416 (1998).

SYNONYM. *Ophrys elatior* Paulus; *O. fuciflora* subsp. *elatior* (Paulus) R.Engel & P.Quentin

DISTRIBUTION. C Europe, centred around the upper Rhine and Rhône valleys.

HABITAT. Full sun to light shade in meadows, grassland, open scrubs and forest edges, usually on calcareous soil; 200–500 m.

FLOWERING. Late June to early September.

DISTINGUISHING FEATURES. 20–90 cm tall. Similar to the typical subspecies but has withered leaves at the peak of flowering, which is remarkably late. The lax, tall inflorescence and small flowers with only 8–10 mm long sepals are also characteristic.

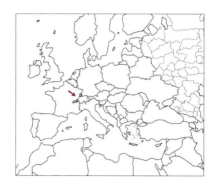

BELOW, LEFT TO RIGHT
subsp. *elatior*
France, Alsace 05.07.2021
France, Alsace 09.07.2021
France, Alsace 05.07.2021
France, Alsace 05.07.2021
France, Alsace 05.07.2021
France, Alsace 09.07.2021

OPPOSITE, LEFT TO RIGHT FROM TOP
subsp. *elatior*
France, Alsace 06.07.2021
France, Alsace 09.07.2021
France, Alsace 09.07.2021
France, Alsace 06.07.2021
France, Alsace 09.07.2021
France, Alsace 09.07.2021
France, Alsace 09.07.2021
France, Alsace 05.07.2021
France, Alsace 09.07.2021

i. *Ophrys holosericea* subsp. *grandiflora*

(H.Fleischm. & Soó) Faurh., *J. Eur. Orch.* 35: 745 (2003)

SYNONYMS. *Ophrys bornmuelleri* f. *grandiflora* H.Fleischm. & Soó; *O. bornmuelleri* subsp. *grandiflora* (H.Fleischm. & Soó) Renz & Taubenheim; *O. levantina* Gölz & H.R.Reinhard; *O. fuciflora* subsp. *grandiflora* (H.Fleischm. & Soó) Faurh., nom. illeg.; *O. levantina* subsp. *grandiflora* (H.Fleischm. & Soó) Kreutz

DISTRIBUTION. Cyprus to S & SE Turkey.

HABITAT. Full sun to light shade in grassland, garigue, maquis and open pine forest, usually on calcareous soil; sea level to 1,400 m.

FLOWERING. Late February to mid-April.

DISTINGUISHING FEATURES. 10–30 cm tall. Similar to subsp. *bornmuelleri* but differs in having an ovate to suborbicular (rarely square) lip with a reduced and often fragmented speculum that is isolated from the lip base or only connected to it by delicate lines. Besides, the lip is always vertically oriented.

BELOW, LEFT TO RIGHT
subsp. *grandiflora*
Cyprus 25.03.1996
Cyprus 15.03.2010
Cyprus 15.03.2003
Cyprus 14.03.2000
Cyprus 15.03.2017
Cyprus 14.03.2000
Cyprus 15.03.1994
Cyprus 12.03.2002

j. *Ophrys holosericea* subsp. *lacaitae* (Lojac.)
W.Rossi, *Inform. Bot. Ital.* 13: 201 (1981, publ. 1983).
SYNONYMS. *Ophrys lacaitae* Lojac.; *O. fuciflora*
subsp. *lacaitae* (Lojac.) Soó; *O. oxyrrhynchos* subsp.
lacaitae (Lojac.) Del Prete.
DISTRIBUTION. Malta (extinct?), Sicily, southern
mainland Italy and Croatia (Vis).
HABITAT. Full sun to light shade in grassland, maquis
and open woodland; sea level to 900 m.
FLOWERING. Late April to June.
DISTINGUISHING FEATURES. 10–35 cm tall. Readily
recognised by its yellow, broadly flabellate lip with
a brown centre around the small speculum. Besides,
the lip is sigmoid in side view.

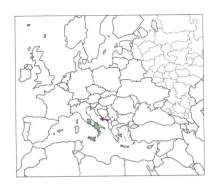

BELOW, LEFT TO RIGHT
subsp. *lacaitae*
Sicily 04.2012 (25)
Sicily 04.2012 (25)
Sicily 04.2012 (25)
Italy, L'Aquila 10.05.2002 (13)
Italy, L'Aquila 04.2012 (25)
Sicily 30.04.2008 (21)

k. *Ophrys holosericea* subsp. *oblita* (Kreutz, Gügel & W.Hahn) H.A.Pedersen & P.J.Cribb, **comb. nov.**

BASIONYM. *Ophrys oblita* Kreutz, Gügel & W.Hahn, *Ber. Arbeitskreis. Heimische Orchid.* 26, 2: 55 (2009).

SYNONYMS. ?*Ophrys episcopalis* var. *samia* P.Delforge; *O. malvasiana* S.Hertel & H.Weyland; ?*O. samia* (P.Delforge) P.Delforge; *O. fuciflora* subsp. *oblita* (Kreutz, Gügel & W.Hahn) Faurh., H.A.Pedersen & S.G.Christ.; ?*O. luminosa* Kreutz, Shifman & M.H.Schot

DISTRIBUTION. From S Greece across S Anatolia to Israel.

HABITAT. Full sun to light shade in grassland, garigue, maquis, olive groves and open pine forest; sea level to 1,700 m.

FLOWERING. April to June.

DISTINGUISHING FEATURES. 30–60 cm, or more, tall. Similar to large-flowered forms of subsp. *holosericea* but distinguished by the flowers being restricted to the upper third to half of the stem and by the lateral margins of the lateral sepals being almost consistently revolute.

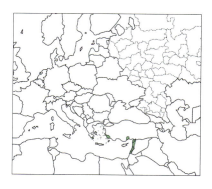

OPPOSITE, FROM TOP LEFT TO RIGHT
subsp. *oblita*
Turkey, Hatay 21.05.2011 (45)
Turkey, Hatay 19.05.2011 (45)
Turkey, Hatay 17.05.2011 (45)
Turkey, Hatay 19.05.2011 (45)
Turkey, Hatay 19.05.2011 (45) – note the lateral sepal margins

l. *Ophrys holosericea* subsp. *oxyrrhynchos* (Tod.) H. Sund., *Taxon* 24: 625 (1975).

SYNONYMS. *Ophrys oxyrrhynchos* Tod.; *O. fuciflora* subsp. *oxyrrhynchos* (Tod.) Soó

DISTRIBUTION. Malta (extinct?), Sicily and S mainland Italy.

HABITAT. Full sun to light shade in grassland, garigue, maquis and open forest on calcareous soil; sea level to 900 m.

FLOWERING. March to May.

DISTINGUISHING FEATURES. 10–30 cm tall. Differs from the other 14 subspecies in having the following combination of features: sepals green to white, at least 4 times as long as the petals; petals about as long as broad; lip flabellate, nearly straight (i.e. not sigmoidly curved in side view), with dark brown ground colour in the central part, occasionally turning light green or yellow towards the margin; terminal appendage more than half as long as the column; speculum covering more than a third of the lip.

NOTE. *Ophrys celiensis* (O.Danesch & E.Danesch) P.Delforge, described from mainland Italy, is morphologically intermediate between *O. holosericea* subsp. *apulica* and subsp. *oxyrrhynchos*; it probably consists of swarms of hybrids between these two taxa.

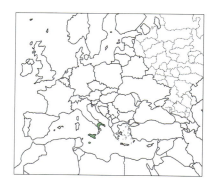

FROM MIDDLE ROW, LEFT TO RIGHT
subsp. *oxyrrhynchos*
Italy, Sicily 11.04.2018
Italy, Sicily 18.04.1999 (45)
Sicily 10.04.2018
Sicily 19.04.2018
Italy, Puglia 22.04.2012 (45)

m. *Ophrys holosericea* subsp. *pallidiconi*

(Faurh.) H.A.Pedersen & P.J.Cribb, **comb. nov.**

BASIONYM. *Ophrys fuciflora* subsp. *pallidiconi* Faurh., *J. Eur. Orch.* 43: 563 (2011).

DISTRIBUTION. SW Turkey (Antalya).

HABITAT. In full sun in garigue and in light shade under scattered cypresses and pines; c. 50 m.

FLOWERING. March and April.

DISTINGUISHING FEATURES. 10–30 cm tall. Reminiscent of subsp. *holosericea* but easily distinguished by the green sepals and the conspicuously white-tipped protuberances of the lip.

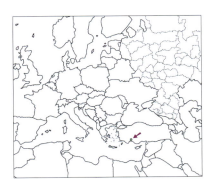

n. *Ophrys holosericea* subsp. *parvimaculata*

(O.Danesch & E.Danesch) O.Danesch & E.Danesch, *Pl. Syst. Evol.* 124: 129 (1975).

SYNONYMS. *Ophrys fuciflora* subsp. *parvimaculata* O.Danesch & E.Danesch; *O. parvimaculata* (O.Danesch & E.Danesch) Paulus & Gack; ?*O. peucetiae* Lozito, D'Emerico, Medagli & Turco

DISTRIBUTION. Endemic to S mainland Italy (Puglia, Basilicata, Campania and Abruzzo).

HABITAT. In open oak (*Quercus pubescens*) and hop hornbeam (*Ostrya carpinifolia*) woods, grassland and *Paliurus spina-christi* scrub; sea level to 600 m.

FLOWERING. April and May.

DISTINGUISHING FEATURES. 10–35 cm tall. Similar to the typical subspecies but differs in having a small, simple, usually H-shaped speculum that is usually less than half as broad as the lip. Besides, the sepals are usually green to white (rarely pink).

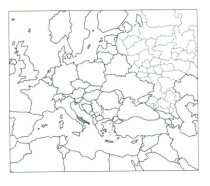

OPPOSITE, LEFT TO RIGHT FROM TOP
subsp. *pallidiconi*
Turkey, Antalya 22.03.2011 (47)
Turkey, Antalya 22.03.2011 (47)
subsp. *parvimaculata*
Italy, Puglia 11.04.2016
Italy, Puglia 10.04.2016
Italy, Puglia 13.04.2016
Italy, Puglia 13.04.2016
Italy, Puglia 14.04.2018
Italy, Puglia 15.04.2016

o. *Ophrys holosericea* subsp. *tetraloniae*

(W.P.Teschner) Kreutz, *Kompend. Eur. Orchid.*: 102 (2004).

SYNONYMS. *Ophrys tetraloniae* W.P.Teschner; *O. posidonia* P.Delforge; *O. fuciflora* subsp. *tetraloniae* (W.P.Teschner) Faurh.

DISTRIBUTION. N Croatia and mainland Italy.

HABITAT. In full sun to light shade in grassland, forest edges and dry hay meadows on calcareous soil; sea level to 800 m.

FLOWERING. June and July.

DISTINGUISHING FEATURES. 30–50 cm tall, rarely shorter. Differs from the other 14 subspecies in having the following combination of features: lowermost flower normally placed 20 cm or more above the ground; flowers small (lip 7–8 mm long); sepals white to green (rarely pink); petals broadly triangular; lip closer to horizontal than vertical orientation, provided with tiny protuberances and a simple speculum.

BELOW, LEFT TO RIGHT
subsp. *tetraloniae*
Croatia, Istria 11.06.2006 (47)
Croatia, Istria 11.06.2006 (47)
Italy, Molise 03.06.2005 (47)

Ophrys scolopax Cav., *Icon.* 2: 46 (1793).
Woodcock orchid

a. subsp. *scolopax*

SYNONYMS. *Ophrys bremifera* Steven; *O. oestrifera* subsp. *bremifera* (Steven) K.Richt.; ?*O. khuzestanica* (Renz & Taubenheim) P.Delforge; *O. santonica* J.M.Mathé & Melki; *O. hygrophila* Gügel, Kreutz, D. & U.Rückbr.; *O. scolopax* var. *minutula* (Gölz & H.R.Reinhard) H.A.Pedersen & Faurh.; *O. karadenizensis* M. & H.Schönfelder; *O. corbariensis* J.Samuel & J.-M.Lewin; *O. ulupinara* W.Hahn, Passin & R.Wegener; *O. scolopax* var. *sepalina* F.M.Vázquez; *O. oestrifera* subsp. *elbursana* Kreutz; *O. mattinatae* Medagli, A.Rossini, Quitadamo, D'Emerico & Turco; ?*O. stavri* (Kalog, Delipetrou & A.Alibertis) P.Delforge

DISTRIBUTION. Most of the Mediterranean region (but very rare in its central part), across Turkey to N Iran and the Caucasus; apparently absent from Morocco, Mallorca, Sardinia, Sicily (except Pantelleria), Crete and Cyprus.

HABITAT. Grassland, garigue, maquis, pine and deciduous woods, olive groves, roadsides and waste places, in full sun to light shade, mostly on calcareous soils; sea level to 2,000 m.

FLOWERING. March to June.

DISTINGUISHING FEATURES. 10–50 cm, or more, tall. Sepals red-purple to white (very rarely green), spreading to reflexed; dorsal sepal from the base at an obtuse angle to the column. Petals triangular to triangular-lanceolate, with spreading hairs.

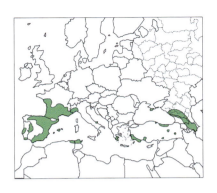

BELOW, LEFT TO RIGHT
O. scolopax subsp. *scolopax*
Turkey, Antalya 25.05.2013 (40)
Turkey, Antalya 16.05.2013 (40)
Turkey, Antalya 16.05.2013 (40)

Lip closer to vertical than horizontal orientation, deeply 3-lobed, usually less than 13 mm long, not constricted at the base, ending in a less than 2.5 mm long (often dentate) appendage that is rectangular, rhombic or obtriangular and narrower than the stigmatic cavity; side lobes converted into obliquely conical, rounded to obtuse (rarely acute) protuberances; speculum consisting of well-defined, often complex marks. Column acute.

NOTE. Southern French and Ligurian populations of plants with a lax inflorescence and a broad lip mid-lobe are often referred to as *O. vetula* Risso and might deserve taxonomic recognition at infraspecific level. However, opinions differ as to the interpretation of *O. vetula*, as some authors believe that the type represents the hybrid *O. holosericea* s.s. × *O. scolopax* s.s. If the latter interpretation proves correct, *O.* ×*vetula* will have to replace *O.* ×*vicina* as the oldest available name for the partly stabilised hybrid complex between *O. holosericea* s.l. and *O. scolopax* s.l..

BELOW, LEFT TO RIGHT
O. scolopax subsp. *scolopax*
France, Aveyron 25.05.2000
France, Cantal 24.05.2017
France, Aveyron 01.06.1995
France, Cantal 25.05.2015
France, Aveyron 22.05.2008
France, Aveyron 22.05.2008
France, Aveyron 18.05.2008
France, Aveyron 22.05.2013

OPPOSITE, LEFT TO RIGHT
subsp. *scolopax*
France, Aveyron 17.05.1999
France, Aveyron 20.05.2002
France, Aveyron 17.05.2011
France, Aveyron 20.05.2002
France, Aveyron 22.05.2007
Turkey, Antalya 23.05.2010 (45)
Spain, Navarre 12.05.2014 (45)
Spain, Navarre 12.05.2014 (45)
France, Aveyron 23.05.2006
France, Aveyron 23.05.2006
France, Aveyron 20.05.2005
France, Aveyron 23.05.2006
Greece, Rhodes 19.04.1996 (45)
France, Aveyron 20.05.2013
France, Aveyron 17.05.2006
France, Aveyron 17.05.2001

b. *Ophrys scolopax* subsp. *apiformis* (Desf.)
Maire & Weiller, *Fl. Afrique N.* 6: 260 (1959,
published 1960).
SYNONYMS. *Ophrys insectifera* var. *apiformis* Desf.;
O. picta Link; *O. sphegifera* Willd.; *O. fuciflora* subsp.
apiformis (Desf.) H.Sund.
DISTRIBUTION. W Mediterranean to Malta.
HABITAT. Full sun to light shade in grassland,
garigue, maquis, olive groves and open pine forest;
sea level to 1,200 m.
FLOWERING. March to June.
DISTINGUISHING FEATURES. 10–40 cm tall. Similar
to small-flowered forms of the typical subspecies
but differs in having linear petals and a lip that is
constricted at the base.

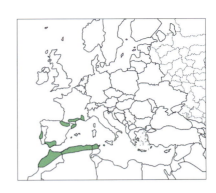

c. *Ophrys scolopax* subsp. *conradiae* (Melki &
Deschâtres) H.Baumann, Giotta, R.Lorenz, Künkele &
Piccitto, *J. Eur. Orch.* 27: 220 (1995).
SYNONYMS. *Ophrys conradiae* Melki & Deschâtres;
?*O. scolopax* subsp. *sardoa* H.Baumann, Giotta,
Künkele, R.Lorenz & Piccitto; *O. scolopax* subsp.
jugurtha R. Martin & Ouni
DISTRIBUTION. Corsica, Sardinia, southern mainland
Italy (Puglia, Basilicata) and possibly Tunisia.
HABITAT. Full sun to light shade in grassland,
garigue, maquis and forest edges; sea level to 400 m.
FLOWERING. Mid-April to mid-June.
DISTINGUISHING FEATURES. 40–65 cm, or less, tall.
Differs from the other eight subspecies in having
the following combination of features: sepals green
or rarely white; petals triangular to triangular-
lanceolate, longer than broad; lip 10–12 mm long,
not constricted at the base; side lobes obliquely
conical, obtuse to rounded, 2–4 mm long; terminal
appendage less than 2.5 mm long, narrower than
the stigmatic cavity; speculum consisting of well-
defined, usually simple marks, normally restricted to
the basal half of the lip.

OPPOSITE, LEFT TO RIGHT FROM TOP
subsp. *apiformis*
France, Aveyron 24.05.2000
France, Aveyron 21.05.2001
France, Aveyron 18.05.1999
France, Aveyron 18.05.2008
France, Aveyron 20.05.2004
France, Aveyron 21.05.2001

subsp. *conradiae*
Sardinia 07.06.1994 (1)
Sardinia 30.05.2010 (21)
Sardinia 07.06.1994 (1)

d. *Ophrys scolopax* subsp. *cornuta* (Steven)
E.G.Camus in E.G.Camus, P.Bergon & A.A.Camus, *Monogr. Orchid.*: 270 (1908).

SYNONYMS. *Ophrys cornuta* Steven; *O. bicornis* Sadler ex Nendtv.; *O. balcanica* Soó; *O. leptomera* P.Delforge; *O. cornutula* Paulus; *O. cerastes* Devillers & Devillers-Tersch.; *O. ceto* Devillers, Devillers-Tersch. & P.Delforge; *O. rhodostephane* Devillers & Devillers-Tersch.; *O. sepioides* Devillers & Devillers-Tersch.; *O. masticorum* P.Delforge & Saliaris; *O. orphanidea* Saliaris & P.Delforge; *O. mycenensis* S.Hertel & Paulus; *O. sappho* Devillers-Tersch., Devillers, Dedroog, Baeten & Flausch; *O. minuscula* (G. & W.Thiele) Presser & S.Hertel; *O. dicipulus* Valahas; *O. cephaloniensis* Paulus

DISTRIBUTION. Italy (Puglia) and E Mediterranean north to Hungary and across to the Crimea and N Iran.

HABITAT. As for the typical subspecies, but almost exclusively on calcareous soil; sea level to 2,000 m.

FLOWERING. March (in the south) to July (in the north).

DISTINGUISHING FEATURES. Plant 20–50 cm tall. Differs from the other eight subspecies in having the following combination of features: sepals rose-purple; lip with 6–12 mm long, horn-like, long-acuminate side lobes.

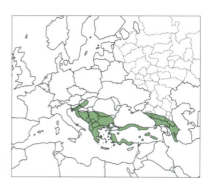

OPPOSITE, LEFT TO RIGHT FROM TOP
subsp. *cornuta*
Greece, Lakonia 26.04.2000 (2)
Rhodes 30.03.2000 (56)
Italy, Calabria 20.04.2010 (21)
Greece, Peloponnese 13.04.2003 (3)
Italy, Calabria 23.04.2008 (21)

subsp. *heldreichii*
Crete 07.04.2014
Crete 07.04.2014
Crete 04.04.2015
Crete 11.04.2015
Crete 06.04.2015

e. *Ophrys scolopax* subsp. *heldreichii* (Schltr.)
E.Nelson, *Gestaltw. Artb. Orchid. Eur. Mittelmeerl.*: 160 (1962).

SYNONYMS. *Ophrys heldreichii* Schltr.; *O. cornuta* subsp. *heldreichii* (Schltr.) Renz; *O. oestrifera* subsp. *heldreichii* (Schltr.) Soó; *O. polyxo* J.Mast, M.-A. Garnier, Devillers-Tersch. & Devillers; *O. hippocratis* P.Delforge

DISTRIBUTION. Apparently endemic to Greece where it is common in Crete, Karpathos and Rhodes but also occurs elsewhere (except in the northernmost regions).

HABITAT. Full sun to light shade in grassland, garigue, maquis and wasteland on calcareous soil; sea level to 1,200 m.

FLOWERING. March to early May.

DISTINGUISHING FEATURES. 15–45 cm tall. Similar to the typical subspecies but differs in having a more than 13 mm long lip with a more than 2.5 mm long terminal appendage.

f. *Ophrys scolopax* subsp. *isaura* (Renz & Taubenheim) H.A.Pedersen & P.J.Cribb in Rolf Kühn et al., *Field Guide Orchids Europe Medit.*: 303 (2019).

SYNONYM. *Ophrys isaura* Renz & Taubenheim.
DISTRIBUTION. CS Turkey (Içel, Antalya).
HABITAT. In full sun to light shade in garigue, scrub and open mixed forest on calcareous ground, usually growing in connection with seeping water; 800–1,200 m.
FLOWERING. May.
DISTINGUISHING FEATURES. 15–50 cm tall. Differs from the other eight subspecies in having the following combination of features: sepals reflexed, light green to dull brownish- or pinkish-green (rarely white); lip with the side lobes transformed into short, obliquely pyramidal protuberances and a much longer slender mid-lobe; apical appendage as wide as, or wider than, the stigmatic cavity.

OPPOSITE PAGE, FIRST THREE ROWS, CLOCKWISE FROM TOP LEFT
subsp. *heldreichii*
Crete 07.04.2014
Crete 10.04.2014
Crete 06.04.2014
Crete 11.04.2004
Crete 12.04.2015
Crete 11.04.2015
Crete 04.04.2015

BOTTOM ROW, LEFT TO RIGHT
subsp. *heldreichii*
Crete 12.04.2015
Crete 06.04.2015
Crete 11.04.2015
subsp. *isaura*
Turkey, Antalya 18.05.2013 (40)
Turkey, Antalya 18.05.2013 (40)
Turkey, Antalya 18.05.2013 (40)
Turkey, Antalya 18.05.2013 (40)
Turkey, Mersin 18.05.1992 (8)

g. *Ophrys scolopax* subsp. *philippi* (Gren.) H.A.Pedersen & P.J.Cribb in Rolf Kühn *et al., Field Guide Orchids Europe Medit.*: 304 (2019).

SYNONYMS. *Ophrys* ×*philippi* Gren., *Mém. Soc. Émul. Doubs* (Sér. 3) 4: 401 (1860); *O. insectifera* subsp. *philippi* (Gren.) Moggr.; *O. scolopax* var. *philippi* (Gren.) Nyman; *O. oestrifera* subsp. *philippi* (Gren.) K.Richt.

DISTRIBUTION. SE France.

HABITAT. In full sun to light shade in grassland, forest edges and garigue; 300–700 m.

FLOWERING. May and June.

DISTINGUISHING FEATURES. 15–35 cm tall. Differs from the other eight subspecies in having the following combination of features: sepals white with green mid-vein; lip generally less vaulted than in the other subspecies; side lobes rounded, widely spreading; speculum covering most of the lip, complicated to marbled.

OPPOSITE, LEFT TO RIGHT FROM TOP
subsp. *philippi*
France, Var 23.05.2010 (50)
France, Var 20.05.2003 (2)
France, Var. 20.05.2003 (2)
France, Var 20.05.2003 (2)
France, Var 20.05.2003
France, Var 23.05.2010 (50)

h. *Ophrys scolopax* subsp. *phrygia* (H.Fleischm. & Bornm.) H.A.Pedersen & P.J.Cribb in Rolf Kühn *et al., Field Guide Orchids Europe Medit.*: 304 (2019).

SYNONYMS. *Ophrys phrygia* H.Fleischm. & Bornm.; *O. oestrifera* subsp. *phrygia* (H.Fleischm. & Bornm.) H.Baumann & R.Lorenz

DISTRIBUTION. S Turkey and NE Aegean Islands (Chios, Lesbos and Evros).

HABITAT. In light shade in overgrown burial grounds, open pine and oak forest, less often in full sun in garigue; sea level to 1,700 m.

FLOWERING. May and June.

DISTINGUISHING FEATURES. 25–80 cm tall. Differs from the other eight subspecies in having the following combination of features: inflorescence very lax, bearing fairly large flowers (lip 10–15 mm long); sepals rose to red-purple (rarely white), reflexed; lip closer to horizontal than vertical orientation; side lobes forming very slender, obliquely conical, distally recurved protuberances that are almost perpendicular to the broadly elliptic mid-lobe.

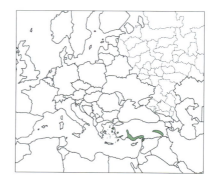

OPPOSITE, LEFT TO RIGHT BOTTOM ROW
subsp. *phrygia*
Turkey, Antalya 13.05.2013 (40)
Turkey, Antalya 13.05.2013 (40)
Turkey, Antalya 14.05.2013 (40)
Turkey, Antalya 14.05.2013 (40)

NOTE. In the southern part of the Turkish province Karaman, plants that differ from typical subsp. *phrygia* in having erect leaves and less consistently slender and perpendicular lip side lobes have been described as *O. kreutzii* W.Hahn, R.Wegener & J.Mast.. The populations concerned are all found at high altitude (1,500–1,700 m) in a vegetation-poor landscape where they occur in patches with seeping water. They might deserve formal recognition at infraspecific level.

BELOW, LEFT TO RIGHT
subsp. *rhodia*
Rhodes 01.04.1999 (22)
Rhodes 05.04.1998 (56)
Rhodes 02.04.2001 (2)
Rhodes 28.03.2000 (51)

i. *Ophrys scolopax* subsp. *rhodia* (H.Baumann & Künkele) H.A.Pedersen & Faurh., *Orchidee (Hamburg)* 48: 235 (1997).

SYNONYMS. *Ophrys umbilicata* subsp. *rhodia* H.Baumann & Künkele; *O. rhodia* (H.Baumann & Künkele) P.Delforge; *O. attica* subsp. *rhodia* (H.Baumann & Künkele) Shifman

DISTRIBUTION. Greece, where it is confined to Rhodes and a couple of smaller neighbouring islands; recurrent reports from Cyprus need verification.

HABITAT. In dry to moist calcareous grassland, garigue, maquis, olive groves and waste places; sea level to 600 m.

FLOWERING. Late March to early May.

DISTINGUISHING FEATURES. 10–35 cm tall. Similar to subsp. *conradiae* but differs in having petals that are approximately as long as broad and a speculum that covers almost the entire lip mid-lobe. Besides, it is usually a shorter plant with a denser inflorescence.

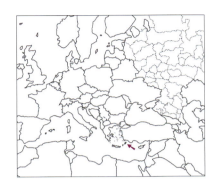

Ophrys umbilicata Desf., *Ann. Mus. Hist. Nat.* 10: 227 (1807). **Eastern woodcock orchid**

a. subsp. *umbilicata*

SYNONYMS. *Ophrys arachnites* var. *attica* Boiss. & Orph.; *O. dinsmorei* Schltr.; *O. attica* (Boiss. & Orph.) Soó; *O. oestrifera* subsp. *orientalis* (Renz) Soó; *O. carmeli* subsp. *attica* (Boiss. & Orph.) Renz; *O. umbilicata* subsp. *beerii* Shifman; ?*O. astarte* Devillers & Devillers-Tersch.

DISTRIBUTION. E Mediterranean from Albania and Greece (excluding Crete) to Iran.

HABITAT. In full sun to light shade in dry to moist calcareous soil in grassland, garigue, open pine and cypress woodland, olive groves and roadside verges; sea level to 1,200 m.

FLOWERING. February to April.

DISTINGUISHING FEATURES. 10–45 cm, or more, tall. Plant slender, usually with erect ovaries. Sepals white to red-purple or green. Lip usually less than 10 mm long and 15 mm broad, not constricted at base; mid-lobe broadest above the middle where the margins are brown (rarely yellow) and completely reflexed, i.e. not visible in front view.

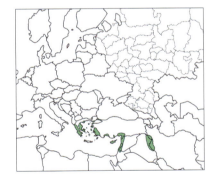

Similar to *O. scolopax* but differs in the petals having appressed (not spreading) hairs and in the dorsal sepal from base being nearly parallel to the column, thus forming a roof over the column.

b. *Ophrys umbilicata* subsp. *bucephala* (Gölz & H.R.Reinhard) Biel, *Ber. Arbeitskreis. Heimische Orchid.* 16 (1): 58 (1999).
SYNONYM. *Ophrys bucephala* Gölz & H.R.Reinhard
DISTRIBUTION. NE Aegean Islands (Lesbos, Chios, Lemnos) and W Turkey.
HABITAT. Garigue, maquis, olive groves and south-facing roadside slopes on calcareous soil; sea level to 500 m.
FLOWERING. Late March to early May.
DISTINGUISHING FEATURES. 10–20 cm tall. Similar to the typical subspecies but differs in having spreading ovaries and a lip that is at least 12 mm long, 15 mm broad and constricted at the base. Besides, the sepals are consistently green to white, occasionally suffused with pink.

BELOW, LEFT TO RIGHT
subsp. *bucephala*
Lesbos 29.03.2018 (45)
Lesbos 29.03.2018 (45)
Lesbos 29.03.2018 (45)
Lesbos 08.04.1998 (1)

OPPOSITE, LEFT TO RIGHT FROM TOP
subsp. *umbilicata*
Cyprus, Troodos 18.03.2002
Cyprus, Troodos 18.03.2002
Cyprus 15.03.2011
Cyprus, Akamus 16.03.2011
Cyprus 18.03.2008
Cyprus 14.03.2009
Cyprus, Akamas 13.03.2011
Cyprus 15.03.2001
Cyprus 16.03.2013
Cyprus 18.03.2008
Cyprus 12.03.2003
Cyprus 14.03.2009

c. *Ophrys umbilicata* subsp. *flavomarginata*
(Renz) Faurh., *J. Eur. Orch.* 35: 745 (2003).
SYNONYMS. *Ophrys attica* f. *flavomarginata* Renz;
O. flavomarginata (Renz) H.Baumann & Künkele;
O. umbilicata subsp. *latilabris* B. & H.Baumann
DISTRIBUTION. Cyprus, Syria to Israel.
HABITAT. Full sun to light shade in grassland, garigue, maquis, olive groves, fallow fields and open pine forest on calcareous soil; sea level to 800 m.
FLOWERING. February to April.
DISTINGUISHING FEATURES. 15–30 cm tall. Similar to the typical subspecies but differs in that the margins of the lip mid-lobe are incompletely reflexed, i.e. visible in front view, and usually yellow (rarely brown). Besides, the sepals are always green and the lip mostly longer than 10 mm.

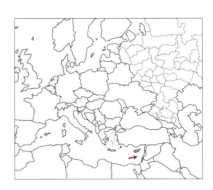

d. *Ophrys umbilicata* subsp. *lapethica* (Gölz & H.R.Reinhard) Faurh., *J. Eur. Orch.* 41: 194 (2009).
SYNONYMS. *Ophrys lapethica* Gölz & H.R.Reinhard;
O. umbilicata subsp. *calycadniensis* Perschke;
O. lapethica subsp. *pamphylica* Kreutz
DISTRIBUTION. Cyprus and CS Anatolia.
HABITAT. Garigue, grassland, fallow fields, olive groves and open pine forest; up to 700 m.
FLOWERING. March and April.
DISTINGUISHING FEATURES. 12–25 cm, or more, tall. Differs from the other four subspecies in having the following combination of features: sepals red-purple, pink, rose, white or green; petals more than half as long as the column; mid-lobe of lip widest around the middle, tapering gradually towards the apex.

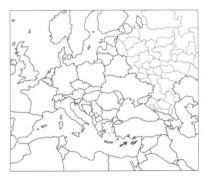

FROM TOP OPPOSITE, LEFT TO RIGHT
subsp. *flavomarginata*
Cyprus 15.03.1999
Cyprus 16.03.2009
Cyprus 17.03.2012
Cyprus 15.03.2011
Cyprus 08.03.1996
Cyprus 16.03.2009
Cyprus 15.03.2011
Cyprus 14.03.2011
Cyprus 19.03.2003
subsp. *lapethica*
Cyprus 12.03.2011
Cyprus 18.03.2000
Cyprus 18.03.2000
Cyprus 18.03.2000
Cyprus 20.03.2011
Cyprus 15.03.2011
Cyprus 17.03.2007
Cyprus 15.03.2009
Cyprus 13.03.2012

e. *Ophrys umbilicata* subsp. *latakiana* (M. & H.Schönfelder) Faurh. & H.A.Pedersen, *J. Eur. Orch.* 40: 695 (2008).

SYNONYMS. *Ophrys latakiana* M. & H.Schönfelder; *O. oestrifera* subsp. *latakiana* (M. & H.Schönfelder) Kreutz

DISTRIBUTION. S Turkey (Hatay) to NW Syria.

HABITAT. In full sun or partial shade in garigue, olive groves and open oak scrubs and on roadside verges; sea level to 800 m.

FLOWERING. April and May.

DISTINGUISHING FEATURES. 30–60 cm tall. Similar to subsp. *lapethica* but differs in that the lowermost flower is placed more than 20 cm above ground and in that the petals are less than half as long as the column. Besides, it is an almost consistently taller plant.

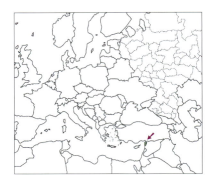

ALL PHOTOS, RIGHT
subsp. *latakiana*
Turkey, Hatay 18.05.2011 (45)

Ophrys cretica (Vierh.) E.Nelson, *Gestaltw. Artb. Orchid. Eur. Mittelmeerl.*: 146 (1962).
Cretan bee orchid

SYNONYMS. *Ophrys spruneri* f. *cretica* Vierh.; *O. cretica* subsp. *karpathensis* E.Nelson, nom. inval.; *O. kotschyi* subsp. *cretica* (Vierh.) H.Sund.; *O. ariadnae* Paulus; *O. cretica* subsp. *ariadnae* (Paulus) H.Kretzschmar; *O. cretica* subsp. *bicornuta* H.Kretzschmar & R.L.Jahn; *O. beloniae* (G. & H.Kretzschmar) Devillers & Devillers-Tersch.

DISTRIBUTION. Southern mainland Greece to the Aegean Islands and Crete.

HABITAT. On dry to moist calcareous soil in full sun or light shade on roadsides and in meadows, garigue, maquis and olive groves; sea level to 1,200 m.

FLOWERING. Mid-February to early May.

DISTINGUISHING FEATURES. 10–40 cm tall. Sepals usually green to brownish. Petals at least 3 times as long as broad. Lip moderately to deeply 3-lobed near the base, with black to dark brown ground colour, ending in a small triangular, more or less porrect point; side lobes converted into hump-shaped or obliquely conical protuberances; speculum highly variable and usually covering most of the mid-lobe, connected to the lip base by distinct broad bands. Stigmatic cavity with white ground colour.

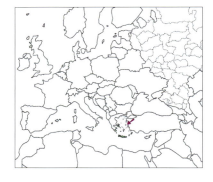

LEFT TO RIGHT
Crete 04.04.2015
Crete 02.04.2014
Crete 04.04.2014
Crete 04.04.2014
Crete 01.04.2014
Crete 10.04.2015

NOTE. This species is often divided into subsp. *cretica* (stigmatic cavity transversely elliptic to circular) and subsp. *ariadnae* (stigmatic cavity rectangular to nearly square), and some authors even split it into two or three separate species. We acknowledge the existence of populations that differ largely in the shape of the stigmatic cavity, but many intermediates occur (questioning the alleged existence of subspecies-specific pollinators), and no other character seems to support the existence of two or three distinct taxa.

BELOW, LEFT TO RIGHT
O. cretica
Crete 01.04.2014
Crete 09.04.2015
Crete 04.04.2015
Crete 11.04.2014
Crete 11.04.2014
Crete 07.04.2014

OPPOSITE, LEFT TO RIGHT, TOP ROW
O. cretica
Crete 04.04.2014
Crete 07.04.2014
Crete 07.04.2014
Crete 07.04.2015

FROM SECOND ROW, TOP TO BOTTOM, LEFT TO RIGHT
Crete 10.04.2015
Crete 04.04.2015
Crete 04.04.2015
Crete 03.04.2014
Crete 04.04.2015
Crete 04.04.2015
Crete 10.04.2015
Crete 08.04.2014

Ophrys kotschyi H.Fleischm. & Soó, *Notizbl. Bot. Gart. Berlin-Dahlem* 9: 908 (1926). **Cyprus bee orchid**

SYNONYMS. *Ophrys sintenisii* subsp. *kotschyi* (H.Fleischm. & Soó) Soó; *O. cypria* Renz

DISTRIBUTION. Cyprus.

HABITAT. In dry to moist grassland, garigue, meadows, olive groves and open pine woods, in full sun to light shade on calcareous soil, or rarely on sandstone; up to 1,000 m.

FLOWERING. Mid-February to early April.

DISTINGUISHING FEATURES. 8–35 cm tall. Similar to *O. cretica* but differs in having petals less than 3 times as long as broad and a stigmatic cavity with dark grey ground colour.

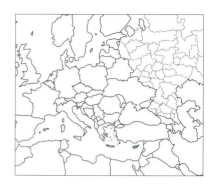

LEFT TO RIGHT
Cyprus 18.03.1996
Cyprus 16.03.2011
Cyprus 16.03.1995
Cyprus 13.03.2011
Cyprus 15.03.1998
Cyprus 15.03.2007

Ophrys reinholdii Spruner ex Fleischm., *Oesterr. Bot. Z.* 57: 5 (1908). **Reinhold's bee orchid**

SYNONYM. *Ophrys spruneri* var. *reinholdii* (Spruner ex Fleischm.) Nyman

a. subsp. *reinholdii*

SYNONYM. *Ophrys reinhardiorum* Paulus

DISTRIBUTION. Balkan Peninsula to SW Turkey.

HABITAT. Garigue, open pine and oak woods, olive groves, roadsides and banks on dry to moist calcareous soils in full sun to light shade; sea level to 1,000 m.

FLOWERING. Early March to early May.

DISTINGUISHING FEATURES. 15–50 cm tall. The species is similar to *O. cretica* but differs in having rose to red-purple (rarely white or pale green) sepals and a speculum that is not connected to the lip base. In subsp. *reinholdii*, the longitudinal bars of the speculum are oblique and usually connected to lateral lip margins, and the column is acute (very rarely acuminate).

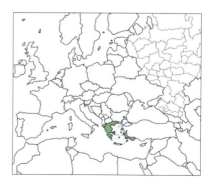

LEFT TO RIGHT
Rhodes 03.04.1988 (40)
Greece, Peloponnese 21.03.2015 (21)
Greece, Chios 16.04.2005 (45)

Turkey, Antalya 18.05.2013 (40)   Turkey, Antalya 18.05.2013 (40)   Turkey, Antalya 12.05.2013 (40)

Turkey, Antalya 12.05.2013 (40)   Turkey, Antalya 29.05.1995 (8)   Turkey, Antalya 16.05.1990 (8)

b. *Ophrys reinholdii* subsp. *straussii*
(H.Fleischm.) E.Nelson, *Gestaltw. Artb. Orchid. Eur. Mittelmeerl.*: 149 (1962). **Strauss's bee orchid**

SYNONYMS. *Ophrys straussii* H.Fleischm.; *O. sintenisii* subsp. *straussii* (H.Fleischm.) Soó; *O. reinholdii* subsp. *leucotaenia* Renz & Taubenheim; *O. antiochiana* H.Baumann & Künkele; *O. straussii* var. *leucotaenia* (Renz & Taubenheim) Ruedi Peter; *O. straussii* subsp. *antiochiana* (H.Baumann & Künkele) Kreutz; *O. reinholdii* subsp. *antiochiana* (H.Baumann & Künkele) H.Baumann & R.Lorenz

DISTRIBUTION. Cyprus (extinct?), CS & SE Turkey, NE Iraq to NW Iran and possibly N Syria.

HABITAT. As for the typical subspecies; 200–1,500 m.

FLOWERING. Late April to early June.

DISTINGUISHING FEATURES. 20–50 cm tall. Distinguished from the typical subspecies in that the longitudinal bars of the speculum are parallel and usually without connection to the lateral lip margins, and in that the column is acuminate (very rarely acute).

NOTE. In 2007, an *Ophrys* colony in and around the overgrown burial ground at Yeniköy in the Turkish province Konya was described as *O. konyana* Kreutz & Ruedi Peter. The colony exhibits a pronounced morphological variation that represents an almost complete continuum between *O. reinholdii* subsp. *straussii* and *O. argolica* subsp. *lucis*. We are probably dealing with a hybrid swarm between these two taxa, both of which have been observed in small numbers at the same site.

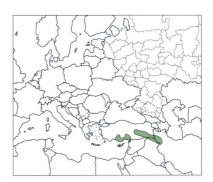

Ophrys cilicica Schltr., *Repert. Spec. Nov. Regni Veg.* 19: 45 (1923). **Taurus bee orchid**

SYNONYMS. *Ophrys kurdica* D. & U.Rückbr.; *O. kurdistanica* Renz

DISTRIBUTION. S and SE Turkey to N Syria, N Iraq and W Iran.

HABITAT. On calcareous soils in full sun to light shade, in grassland, scrub and open forest and amongst rocks; 500–1,400 m.

FLOWERING. April to May.

DISTINGUISHING FEATURES. 15–40 cm, or more, tall. Similar to *O. cretica* but easily recognised by its much more slender habit, its narrow, stalk-like lip base and its narrow stigmatic cavity.

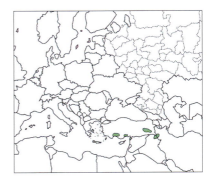

CLOCKWISE FROM FAR LEFT
Turkey, Antalya 16.05.2013 (40)
Turkey, Mersin 10.05.1989 (25)
Turkey, Diyarbakir 19.05.1974
Turkey, Diyabakir 08.05.1974 (10)

PARTLY STABILISED HYBRID COMPLEXES

The following common and often widespread natural hybrids are included here as each is frequently treated as one or more species in other recent treatments. We informally recognise morphologically distinct elements within some of these partly stabilised hybrid complexes as nothomorphs. These probably originate from the hybridisation of different subspecies of the respective parental species. Often, the notomorphs coincide with taxa recognised by other authors at species rank or below. However, their hybrid origin, whether recent or historic, is usually morphologically clear. The boundaries of the notomorphs are often difficult to determine because of the inclusion in each of primary (F1) and later (F2, F3 etc.) crosses as well as back-crosses to either parent. Less frequent and occasional hybrids are treated at the end of the species accounts.

Ophrys ×*brigittae* H.Baumann, *Beih. Veröff. Naturschutz Landespflege Baden-Württemberg* 19: 142 (1981).

PARENTAGE. *Ophrys fusca* × *O. omegaifera*

DISTRIBUTION. SW France to Portugal, E mainland Spain and the Balearic Islands in the W Mediterranean and from SW Turkey to the SE Aegean Islands and Crete.

HABITAT. Full sun to light shade in grassland, garigue, maquis and open pine forest, mostly on calcareous soil; sea level to 1,100 m.

FLOWERING. January to May.

DISTINGUISHING FEATURES. 7–20 cm, or more, tall. Largely intermediate between *O. fusca* and *O. omegaifera*. It is mainly differentiated from both by the obscure longitudinal furrow at the base of its lip (the furrow is distinct in *O. fusca* and lacking in *O. omegaifera*).

NOTE. The forms that make up this taxon probably constitute a partly stabilised hybrid complex between *O. fusca* and *O. omegaifera*, with at least two subspecies of the latter being involved (subsp. *dyris* in the west, subsp. *omegaifera* in the east). Two informal nothomorphs can be characterised:

Nm. "*vasconica*" has a shallowly 3-lobed lip with a usually unmarbled speculum. W Mediterranean.
SYNONYM. *O. vasconica* (O. & E.Danesch) P.Delforge

Nm. "*sitiaca*" has a moderately 3-lobed lip with a usually marbled speculum. E Mediterranean.
SYNONYMS. *Ophrys creutzburgii* H.Baumann & Künkele; *O. sitiaca* Paulus, C. & A.Alibertis; ?*O. pelinaea* P.Delforge; *O. scyria* P.Delforge & Onckelinx

OPPOSITE PAGE, LEFT TO RIGHT FROM TOP
Nm. '*vasconica*'
Spain, Álava (45)
Spain, Álava (45)
Spain, Navarre (45)

Nm. '*sitiaca*'
Turkey, Antalya (2)
Greece, Rhodes (53)
Greece, Rhodes (53)

Ophrys* ×*flavicans Vis., *Fl. Dalmat.* 1: 178 (1842).

PARENTAGE. *Ophrys bertolonii* × *O. sphegodes*

DISTRIBUTION. Southern Europe from NE Spain to Malta and Dalmatia; absent from Corsica and Sardinia.

HABITAT. Full sun to light shade in grassland, garrigue, open maquis, olive groves and open patches in forest, mainly on calcareous soil; sea level to 1,300 m.

FLOWERING. March to June.

DISTINGUISHING FEATURES. 9–35 cm, or more, tall. Largely intermediate between *O. bertolonii* and *O. sphegodes*. Distinguished from the former by its slightly vaulted to only moderately saddle-shaped

lip and in that its column does not taper towards the base (in side view). Distinguished from *O. sphegodes* mainly by its speculum being almost consistently isolated from the lip base.

NOTE. The multitude of forms making up this taxon all seem to be part of a partly stabilised hybrid complex between *O. bertolonii* and *O. sphegodes*, the latter possibly involving more than one subspecies. Ten informal nothomorphs can be characterised:

Nm. "benacensis" has usually pink sepals and a lip that is 15–21 mm long, straight to moderately saddle-shaped, usually entire, devoid of protuberances, dark brown with (red-)brown hairs in the marginal zone and emarginate around an upcurved apical point. The speculum is usually coherent and placed above the middle of the lip. SE France and N Italy.

SYNONYMS. *Ophrys saratoi* E.G.Camus; *O. pseudobertolonii* Murr; *O. benacensis* (Reisigl) O. & E.Danesch; *O. aurelia* P.Delforge

Nm. "catalaunica" is similar to nm. "benacensis", but differs in having generally smaller flowers (lip 9–16 mm long) and in that the lip is slightly vaulted and apically rounded with a porrect point. NE Spain and SW France.

SYNONYMS. *Ophrys catalaunica* O. & E.Danesch; *O. magniflora* Geniez & Melki

Nm. "balearica" is similar to nm. "benacensis", but is distinguished by its always straight and usually 3-lobed lip. Balearic Islands.

SYNONYM. *Ophrys balearica* P.Delforge

Nm. "drumana" has rose to pink or rarely white sepals and a lip that is 9–12 mm long, straight, usually entire, devoid of protuberances, dark brown with (red-)brown hairs in the marginal zone and apically rounded with an upcurved point. The speculum is coherent and placed above the middle of the lip. SE France.

SYNONYM. *Ophrys drumana* P.Delforge

Nm. "flavicans" is similar to nm. "benacensis", but is distinguished by its smaller flowers (lip under 12 mm long) and by its lip having yellow-brown hairs in the marginal zone. C Dalmatia.

OPPOSITE PAGE, LEFT TO RIGHT FROM TOP
Nm. 'benacensis'
Italy, Verona (2)
Italy, Verona (11)
Italy, Como (11)

Nm. 'catalaunica'
Spain, Catalunya (2)
France, Languedoc (13)
France, Languedoc (13)

Nm. 'balearica'
Majorca
Majorca
Majorca

ABOVE, FROM TOP
Nm. 'drumana'
France (50)
France, Drome (2)
France, Drome (2)

Nm. **"bertoloniiformis"** is similar to nm. *"benacensis"*, but is distinguished by its smaller flowers (lip under 15 mm long) with almost consistently green sepals. Southern Italy.

SYNONYM. *Ophrys bertoloniiformis* O. & E.Danesch

Nm. **"promontorii"** has green sepals and a lip that is 10–15 mm long, straight, entire with low protuberances, dark brown with (red-)brown hairs in the marginal zone and apically rounded with a porrect point. The speculum is normally fragmented. Southern mainland Italy.

SYNONYM. *Ophrys promontorii* O. & E.Danesch

Nm. **"tarentina"** is similar to nm. *"promontorii"*, but differs in its light brown lip with a usually coherent speculum. Southern mainland Italy.

SYNONYM. *Ophrys tarentina* Gölz & H.R.Reinhard

Nm. **"explanata"** is similar to nm. *"benacensis"*, but is distinguished by its smaller flowers (lip under 15 mm long) and in its often 3-lobed lip. Sicily.

SYNONYM. *Ophrys explanata* (Lojac.) P.Delforge

Nm. **"melitensis"** has green (rarely white to pale rose) sepals and a lip that is 12–16 mm long, straight, entire, devoid of protuberances, dark brown with (red-)brown hairs in the marginal zone and emarginate around a usually porrect apical point. The speculum is coherent and usually placed below the middle of the lip. Malta.

SYNONYM. *Ophrys melitensis* (Salk.) Devillers-Tersch. & Devillers

LEFT, TOP TO BOTTOM
Nm. '*explanata*' Sicily (3)
Nm. '*melitensis*' Malta (3)

OPPOSITE, FROM TOP TO BOTTOM, LEFT COLUMN
Nm. '*bertoloniiformis*'
Italy, Puglia
Italy, Puglia
Italy, Puglia

MIDDLE COLUMN
Nm. '*promontorii*'
Italy, Puglia (9)
Italy, Puglia
Italy, Puglia (9)

RIGHT COLUMN
Nm. '*tarentina*'
Italy, Puglia
Italy, Puglia
Italy, Puglia
Italy, Puglia
Italy, Puglia

Ophrys* ×*arachnitiformis Gren. & Philippe, *Mém. Soc. Émul. Doubs* (Sér. 3) 4: 391 (1860).

PARENTAGE. *Ophrys holosericea* × *O. sphegodes*

SYNONYMS. *Ophrys morisii* (Martelli) G.Keller & Soó; *O. castellana* Devillers-Tersch. & Devillers; *O. panattensis* Scrugli, Cogoni & Pessei

DISTRIBUTION. Southern Europe from NE Spain to Dalmatia and southern mainland Italy; absent from the Balearic Islands and Sicily, also an isolated ocurrence in N France.

HABITAT. Full sun to moderate shade in grassland, garigue, maquis, wooded meadows and open forest and on roadside slopes; sea level to 1,000 m.

FLOWERING. February to early June.

DISTINGUISHING FEATURES. 10–40 cm, or more, tall. Somewhat intermediate between *O. holosericea* and *O. sphegodes*, but closer to the latter and not likely to be confused with *O. holosericea*. Mainly distinguished from *O. sphegodes* by its more (though not densely) hairy petals and by generally higher levels of within-population variation in sepal colour, lip shape and speculum pattern.

NOTE. The multitude of forms that make up this taxon all seem to be part of a partly stabilised hybrid complex between *O. holosericea* and *O. sphegodes*, both possibly involving more than one subspecies. Four informal nothomorphs can be characterised (but do not take into account all populations):

Nm. "*arachnitiformis*" has white to rose (rarely green) sepals and 2–3 mm broad petals with flat or wavy margins; the lip is elliptic and the speculum usually not delimited by a white border. Early-flowering, February to mid-April. NE Spain to NW Italy (Liguria).

SYNONYMS. *Ophrys sphegodes* subsp. *integra* (Moggr. & Rchb.f.) H.Baumann & Künkele; *O. exaltata* subsp. *marzuola* Geniez, Melki & Soca

Nm. "*splendida*" has white to rose sepals and 3–5 mm broad petals with wavy margins; the lip is elliptic and the speculum always delimited by a conspicuous white border. Late-flowering, mid-April to early June. SE France to Liguria.

SYNONYM. *Ophrys splendida* Gölz & H.R.Reinhard

Nm. "*montis-leonis*" has white to red-purple sepals and 3–5 mm broad petals with flat margins; the lip is normally trapeziform and the speculum usually delimited by a narrow white border.

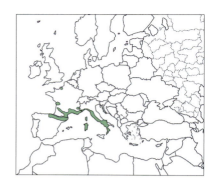

OPPOSITE, LEFT TO RIGHT FROM TOP
O. ×*arachnitiformis*
Sardinia 16.04.2016
Sardinia 09.04.2016
Nm. '*arachtiniformis*'
France, Provence (13)
Nm. '*splendida*'
France, Provence
France, Provence
France, Provence
France, Provence
France, Provence
Nm. '*montis-leonis*'
Italy, Tuscany (56)
Italy, Tuscany (56)
Italy, Tuscany (56)
Nm. '*archipelagi*'
Italy, Puglia (2)
Italy, Puglia (45)

Flowering March to May. WC Italy.

SYNONYMS. *Ophrys montis-leonis* O. & E.Danesch; *O. tyrrhena* Gölz & H.R.Reinhard

Nm. "*archipelagi*" has yellowish-green to whitish-green sepals and 3–5 mm broad petals with wavy margins; the lip is elliptic and the speculum usually delimited by a narrow white border. Flowering March to May. SC Italy and Dalmatia.

SYNONYM. *Ophrys archipelagi* Gölz & H.R. Reinhard

Ophrys ×delphinensis O. & E.Danesch, *Orch. Eur. Ophrys-Hybr.*: 222 (1972).

PARENTAGE. *Ophrys argolica × O. scolopax*

DISTRIBUTION. Endemic to Greece where it is centred at the Corinthian Gulf.

HABITAT. Full sun to light shade in meadows, maquis and clearings on calcareous soil; sea level to 1,300 m.

FLOWERING. April and May.

DISTINGUISHING FEATURES. 20–40 cm tall. Largely intermediate between *O. argolica* subsp. *argolica* and *O. scolopax* s.l. The conspicuous masses of long, whitish hairs on the basal part of the lip readily separate this complex from genetically pure *O. scolopax*, whereas variation in the lobing of the lip and in the size of its protuberances often challenges separation from *O. argolica*.

NOTE. The forms that make up this taxon probably constitute a partly stabilised hybrid complex between *O. argolica* subsp. *argolica* and one of the co-occurring subspecies of *O. scolopax*.

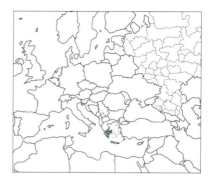

BELOW, LEFT TO RIGHT
O. ×delphinensis
Greece, Achaea (2)
Greece, Attica (11)
Greece, Attica (11)
Greece, Arachova (11)
Greece, Ano Diakopto (11)
Greece, Arachova (2)

Ophrys* ×*vicina Duffort, *Orch. Gers.*: 26 (1902).
PARENTAGE. *Ophrys holosericea* × *O. scolopax*
SYNONYM. *Ophrys homeri* M.Hirth & H.Spaeth
DISTRIBUTION. SW Turkey, Aegean Islands and SE Europe (NE Italy, Slovakia, Hungary and ?Croatia).
HABITAT. Full sun to light shade in grassland, garigue, open maquis and pine forest on calcareous soil; sea level to 700 m.
FLOWERING. Early March (in the south) to June (in the north).
DISTINGUISHING FEATURES. 7–40 cm, or more, tall. Largely intermediate between *O. holosericea* and *O. scolopax*, but distinguished from both mainly by its higher within-population variation in degree of lobing of the lip and in the size and position of the lip protuberances. Furthermore, *holosericea*-like and *scolopax*-like flowers are sometimes borne in the same inflorescence.
NOTE. The multitude of forms that make up this taxon all seem to be part of a partly stabilised hybrid complex between *O. holosericea* and *O. scolopax*, both possibly involving more than one subspecies. Three informal nothomorphs can be characterised (but do not take into account all populations):

Nm. "*calypsus*" has large flowers (petals at least 5 mm long, lip at least 12 mm broad) with spreading sepals and a strongly vaulted lip, the protuberances of which are straight to slightly curved and usually over 2.5 mm long. Aegean Islands to SW Turkey.
SYNONYMS. *Ophrys calypsus* M.Hirth & H.Spaeth; *O. heldreichii* var. *pseudoapulica* P.Delforge; *O. heldreichii* var. *scolopaxoides* P.Delforge

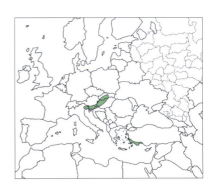

BELOW, ALL IMAGES
O. ×*vicina*
Nm. '*calypsus*'
Rhodes (56)

Nm. "*heterochila*" has small flowers (petals less than 4 mm long, lip less than 12 mm broad) with reflexed petals and a slightly vaulted lip, the protuberances of which are straight to distinctly curved and usually under 2.5 mm long. Sporades to SW Turkey.

SYNONYMS. *Ophrys heterochila* (Renz & Taubenheim) Paulus & Gack; *O. dodekanensis* H.Kretzschmar & Kreutz; *O. oreas* Devillers & Devillers-Tersch.; *O. chalkae* M.Hirth & H.Spaeth; *O. tili* M.Hirth & H.Spaeth; *O. ellinicaea* Kreutz & Tenschert; *O. samiotissa* M.Hirth & Paulus

Nm. "*holubyana*" has medium-sized flowers (petals 3–6 mm long, lip 10–17 mm broad) with spreading sepals and a moderately vaulted lip, the protuberances of which are distinctly curved and 2–8 mm long. NE Italy to the Carpathians.

SYNONYMS. *Ophrys holubyana* András.; *O. zinsmeisteri* A.Fuchs

LEFT TO RIGHT
O. ×*vicina*
Nm. '*heterochila*'
Greece, Chios (45)
Greece, Chios (45)
Rhodes (56)
Nm. '*holubyana*'
Italy (11)

27. STEVENIELLA Schltr., *Repert. Spec. Nov. Regni Veg.* 15: 292 (1918).

A monotypic genus from the Crimea, Caucasus, Turkey and Iran.

Steveniella satyrioides (Spreng.) Schltr., *Repert. Spec. Nov. Regni Veg.* 15: 295 (1918). **Steven's orchid**

SYNONYMS. *Himantoglossum satyrioides* Spreng.; *Stevenorchis satyrioides* (Spreng.) Wankow & Kraenzl.

DISTRIBUTION. N Turkey, Crimea and Caucasus, east to NW Iran.

HABITAT. Somewhat shady places in deciduous and coniferous woods and forests, occasionally in marshes and grassland; up to 2,100 m.

FLOWERING. April and May.

DISTINGUISHING FEATURES. 25–40 cm tall. A distinctive species with a single green or purplish leaf at the base and bicoloured flowers with a trilobed lip in which the mid-lobe is oblong and much longer than the triangular side lobes, whereas the spur is partly cleft.

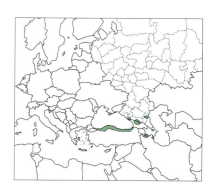

ALL IMAGES
Georgia 17.04.2013 (27)

338 HIMANTOGLOSSUM

28. HIMANTOGLOSSUM Spreng., *Syst. Veg.* (Ed. 16) 3: 675, 694 (1826).

SYNONYMS. *Loroglossum* Rich.; *Comperia* K. Koch; *Barlia* Parl.

A genus of eleven species ranging from the Canary Islands, NW Africa and the Iberian peninsula across to Iran. Most species are subtropical, but *H. hircinum* extends northwards to Central Europe, western France and England.

These are large and coarse plants with distinctive, showy flowers. In all species but one, the side lobes of the lip have distinctly wavy margins, and in the only species where this is not the case, the lip ends in four thread-like processes. Most species have a conspicuously prolonged lip midlobe, and some of them have flowers that smell of goats. Identification key on p. 432.

Himantoglossum comperianum (Steven)
P. Delforge, *Naturalistes Belges* 80 (3): 401 (1999).
Comper's orchid

SYNONYMS. *Orchis comperiana* Steven; *Comperia comperiana* (Steven) Asch. & Graebn.

DISTRIBUTION. Dodecanese, Rhodes, the Crimea, and Turkey south to the Lebanon and east to NW Iran.

HABITAT. On forest edges and bushy scrub, often in burial grounds, on limestone soils; 400 to 2,000 m.

FLOWERING. May to July.

DISTINGUISHING FEATURES. 25–65 cm tall. Its large flowers are very distinctive, with a pink-purple, unmarked lip, which is fan-shaped at the base and with long slender tails to each lobe.

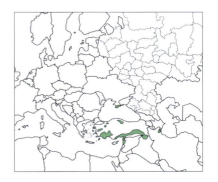

CLOCKWISE FROM TOP LEFT
Turkey, Antalya 12.05.2013 (40)
Turkey, Antalya 12.05.2013 (40)
Turkey, Antalya 25.05.2013 (40)
Iran, Kermanshah 08.05.2000 (2)
Turkey, Ceviz 15.05.1988 (10)
Turkey, Antalya 17.05.2013 (40)

Himantoglossum robertianum (Loisel.)
P. Delforge, *Naturalistes Belges* 80 (3): 401 (1999).
Giant orchid

SYNONYMS. *Orchis robertiana* Loisel., *Barlia robertiana* (Loisel.) Greuter

DISTRIBUTION. Temperate France, Mediterranean Europe, Cyprus and Turkey; also in Mediterranean N Africa.

HABITAT. Sunny to semi-shaded places in coastal and hill country, in garigue, open forest, sparse grassland and bush; sea level to 1,700 m.

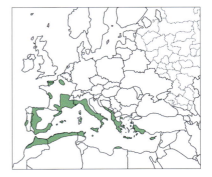

LEFT TO RIGHT
Cyprus 15.03.2002
Cyprus 18.03.1999
Cyprus 18.03.1996

FLOWERING. February to April.

DISTINGUISHING FEATURES. 30–80 cm tall. A distinctive early-flowering orchid with all leaves assembled in a basal rosette. The large, often dull-coloured flowers have a trilobed lip with short somewhat tapering side lobes and a somewhat longer mid-lobe that is deeply emarginated or bilobed at the tip. The lip varies from white to purple in ground colour, usually with more or less linear markings and often with an olive-green marginal zone.

BELOW, LEFT TO RIGHT
H. robertianum
Cyprus 15.03.2002
Cyprus 18.03.1999
Cyprus 18.03.1999
Cyprus 21.03.2001
Italy, Puglia 15.03.2009 (21)

Himantoglossum metlesicsianum (W.P. Teschner) P. Delforge, *Naturalistes Belges* 80 (3): 401 (1999).
Tenerife giant orchid

SYNONYM. *Barlia metlesicsiana* W.P. Teschner

DISTRIBUTION. Endemic to Tenerife. An endangered species.

HABITAT. Full sun to partial shade in garrigue and abandoned agricultural terraces on crumbled lava; 400 to 1,200 m.

FLOWERING. December to February.

DISTINGUISHING FEATURES. 30–80 cm tall. Differs from the widespread *H. robertianum* in having leaves that are distributed along the stem and a pink lip with a pale centre, lacking any green colouration.

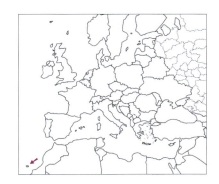

CLOCKWISE FROM LEFT
Tenerife 13.02.2003 (3)
Tenerife 13.01.2003 (3)
Tenerife 02.1996 (22)

Himantoglossum formosum (Stev.) Koch, *Linnaea* 22: 287 (1849).

SYNONYMS. *Orchis formosa* Stev.; *Loroglossum formosum* (Stev.) E.G.Camus

DISTRIBUTION. E Caucasus, Azerbaijan and NW Iran.

HABITAT. In light shade in scrub and oak woodland on moist calcareous soils; up to 700 m.

FLOWERING. May and June.

DISTINGUISHING FEATURES. Flowers unspotted; lip 24–30 mm long, rich purple with a pale disc, side lobes short with undulate margins, the mid-lobe linear, notched at the tip, twisted in the apical part, 13–17 mm long; spur slender, 7–10 mm long.

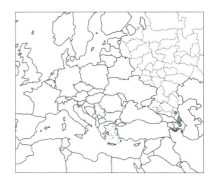

Himantoglossum hircinum (L.) Spreng., *Syst. Veg.* (Ed. 16) 3: 694 (1826). **Lizard orchid**

SYNONYMS. *Satyrium hircinum* L.; *Aceras hircinum* (L.) Lindl

DISTRIBUTION. W and C Europe and W Mediterranean (absent from Ireland, the Nordic countries and Portugal and SE and E Europe); also in NW Africa.

HABITAT. Open forest, bush and meadows, on calcareous soil; sea level to 1,800 m.

FLOWERING. May to August.

DISTINGUISHING FEATURES. 30–120 cm tall. Distinguished from the other species of lizard orchid by its entire or rather shortly notched (up to 5 mm) mid-lobe, slender side lobes that are 13–18.5 mm long and a spur that is 2–3 mm long.

FROM TOP LEFT TO RIGHT
France, Alsace 28.05.2021
France, Alsace 28.05.2021
France, Jura 25.05.2019 (25)
France, Aveyron 28.05.2001
France, Haute Provence 12.05.2005
France, Languedoc-Roussillon 17.05.2009
France, Haute Provence 12.05.2009
France, Languedoc-Roussillon 17.05.2009

HIMANTOGLOSSUM | 343

Azerbaijan, Kuba
28.05.1997 (1)

Azerbaijan, Kuba
28.05.1997 (1)

Azerbaijan, Kuba
03.06.1994 (5)

Himantoglossum adriaticum H.Baumann, *Orchidee (Hamburg)* 29 (4): 171 (1978).
Adriatic lizard orchid

DISTRIBUTION. From mainland Italy across Croatia, Slovenia and Austria to Hungary, Slovakia and the Czech Republic.

HABITAT. Open forest, bush, and meadows, on calcareous soil; sea level to 1,900 m.

FLOWERING. May to July.

DISTINGUISHING FEATURES. 30–80 cm tall. Closely allied to *H. hircinum* but differs in its much laxer inflorescence with flowers that have a narrow column and a lip with narrow shoulders, darker red-purple markings and a deeply notched (5–18 mm) tip to the lip.

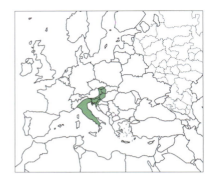

BELOW, LEFT TO RIGHT
Italy, Abruzzo 28.05.1998 (1)
Italy, Abruzzo 29.06.2004 (1)
Italy, Umbria 01.06.2008 (21)
Italy, Umbria 28.05.2008 (21)
Italy, Umbria 23.05.2008 (21)

Himantoglossum caprinum (M. Bieb.) Spreng., *Syst. Veg.* (Ed. 16) 3: 694 (1826).

SYNONYMS. *Orchis caprina* M.Bieb.; *Himantoglossum hircinum* var. *caprinum* (M.Bieb.) W. Zimm.; *H. affine* (Boiss.) Schltr.; *H. bolleanum* (Siehe) Schltr.; *H. caprinum* subsp. *bolleanum* (Siehe) H.Baumann & R.Lorenz; *H. caprinum* subsp. *levantinum* B. & H. Baumann; *H. galilaeum* Shifman

DISTRIBUTION. Crimea, Turkey to Israel and N Iran.

HABITAT. Open oak forests, forest margins, bush and meadows on calcareous soil; sea level to 1,500 m.

FLOWERING. May to August.

DISTINGUISHING FEATURES. 25–80 cm tall. Inflorescence lax, 10–30-flowered. Differs from *H. hircinum* in having 6–9 mm wide, ovate-lanceolate sepals; an unspotted lip with a hairy centre; a 45–68 mm long deeply divided mid-lobe; linear, incurved, 10–30 mm long side lobes; and a 4–5 mm long spur. It differs from *H. adriaticum* in having flowers with 13–18.5 mm long lateral sepals (vs. 8–11.5 mm). Closely allied to *H. calcaratum* and *H. montis-tauri*.

NOTE. *Himantoglossum samariense* C. & A.Alibertis from Crete and the Peloponnese is now considered to consist of swarms of hybrids between *H. caprinum* and *H. calcaratum* s.l..

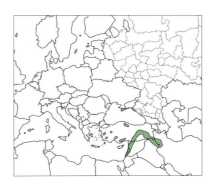

BELOW, LEFT TO RIGHT
Turkey, Verziköprü 11.06.1989 (1)
E Turkey 15.05.1996 (21)
Turkey, Konya 26.05.2010 (45)

Himantoglossum calcaratum (G. Beck) Schltr., Repert. Spec. Nov. Regni Veg. Sonderbeih. A 1: 145 (1927).

SYNONYMS. *Himantoglossum hircinum* subsp. *calcaratum* (G. Beck) Soó; *H. jankae* var. *calcaratum* (Beck) P.Delforge

a. subsp. *calcaratum*

DISTRIBUTION. Albania and former Yugoslavia.

HABITAT. Meadows, the edges of thickets and roadsides; 400 to 1,000 m.

FLOWERING. June to early August.

DISTINGUISHING FEATURES. Differs from *H. hircinum* in being more robust, with 8–10, broadly lanceolate leaves and a denser inflorescence of flowers with a larger, longer lip with short incurved side lobes that have undulate margins, a longer (75–100 mm) mid-lobe that is deeply bifid for 15–25 mm at the tip, and an 8–12 mm long spur. It differs from subsp. *rumelicum* in being a more robust plant that has a lip with longer segments and a longer spur. The relationship between the typical subspecies and subsp. *rumelicum* requires further investigation.

OPPOSITE, CLOCKWISE FROM TOP LEFT
subsp. *calcaratum*
Kosovo, Pec 06.07.1995 (1)
Bosnia, Sutjeska valley 05.07.2015 (58)
Kosovo, Laliquan 02.07.2015 (58)
Serbia, Nova Varop 10.08.2014 (58)

b. *Himatoglossum calcaratum* subsp. *rumelicum* (H.Baumann & R.Lorenz) Niketic & Djordevic, Bull. Nat. Hist. Mus. Belgrade 11: 104 (2018).

SYNONYMS. *Himantoglossum caprinum* auct.; *H. caprinum* subsp. *rumelicum* H.Baumann & R.Lorenz; *H. caprinum* subsp. *robustissimum* Kreutz; *H. jankae* Somlyay, Kreutz & Óvári; *H. jankae* var. *rumelicum* (H.Baumann & R.Lorenz) P.Delforge; *H. calcaratum* subsp. *jankae* (Somlyay, Kreutz & Óvári) R.M.Bateman, Kreutz & Óvári

DISTRIBUTION. E Europe from Czech Republic, Hungary and SE Europe across to Greece and Turkey.

HABITAT. Dry grassland, forest edges, scrubland, on calcareous soil.

FLOWERING. June to August.

DISTINGUISHING FEATURES. 30–100 cm tall. Distinguished by its relatively large reddish flowers, which have 13–18.5 mm long lateral sepals, a 46–90 mm long lip with red papillate spots on the basal area and a relatively long spur (6.5–9 mm long). However, it is exceedingly variable and more work is needed to clarify its status.

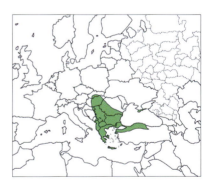

OPPOSITE, LEFT TO RIGHT FROM BOTTOM LEFT
subsp. *rumelicum*
Greece, Pindos, Ioannina 14.06.2018 (59)
Greece, Pindos, Ioannina 14.06.2018 (59)
Greece, Pindos, Ioannina 14.06.2018 (59)
Greece, Pindos, Ioannina 14.06.2018 (59)

Himantoglossum montis-tauri Kreutz & W. Lüders, *J. Eur. Orch.* 29 (4): 655 (1997).
Taurus lizard orchid
DISTRIBUTION. SW Turkey; records from Lesbos probably represent mis-identifications of *H. calcaratum* subsp. *rumelicum*.
HABITAT. Open oak and coniferous forest, garigue, on calcareous soil; 800 to 1,200 m.
FLOWERING. Mid-May to late June.
DISTINGUISHING FEATURES. 20–90 cm tall. It has fewer leaves than *H. caprinum* and larger floral bracts, twice as long as the ovary. Inflorescence 8–30-flowered. Lip with a 40–70 mm long mid-lobe, deeply bifid in the apical fifth, reflexed, with 7–10 mm long side lobes and a 3–4 mm long spur. Centre of the lip white with purple spots (often in the shape of transverse bars), not with tufts of short purple hairs. Close to *H. calcaratum* subsp. *rumelicum* and possibly should be included there.

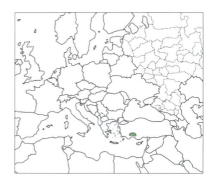

29. SERAPIAS L., *Sp. Pl.*: 949 (1753).

SYNONYMS. *Isias* De Not; *Serapiastrum* Kuntze; *Lonchitis* Bubani
DISTRIBUTION. Throughout the Mediterranean region, with extensions to Macaronesia, Western Europe, the southern Alps and the Caucasus.
10 species. Readily recognised by the large, striated bracts that partly envelope the flowers and by the lip, which consists of a hypochile with incurved side lobes (more or less hidden by the connivent sepals) and a flat, usually hairy epichile. Identification key on p. 433.

Serapias cordigera L., *Sp. Pl.* (Ed. 2): 1345 (1763).
Heart-lipped tongue-orchid
SYNONYMS. *Serapias azorica* Schltr.; *S. cossyrensis* B.Baumann & H.Baumann; *S. cordigera* var. *cretica* (B.Baumann & H.Baumann) P.Delforge; *S. cordigera* var. *mauritanica* (E.G.Camus) E.Nelson ex P.Delforge
DISTRIBUTION. From the Azores across the Mediterranean region to W Turkey; reaching Austria in the north.
HABITAT. Mainly occurring in garigue, meadows and open forest, but can also be found on rocky slopes, abandoned agricultural terraces and in open places in maquis, always on acid soil; sea level to 1,000 m.
FLOWERING. March to June.
DISTINGUISHING FEATURES. 10–40 cm, or more, tall. Inflorescence dense at peak flowering (any bract in the mid-portion of the spike overlapping with the

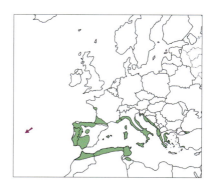

Turkey, Antalya
22.05.2013 (40)

Turkey, Antalya
22.05.2013 (40)

Turkey, Antalya
22.05.2013 (40)

Turkey, Antalya
22.05.2013 (40)

France, Haute-Provence
12.05.1995

France, Haute-Provence
12.05.2004

France, Haute-Provence
09.05.2005

bract immediately above for more than half of its length). Flowers allogamous (ovaries usually not swelling while the perianths are still fresh). Lip at the base with two divergent keels; epichile 14–29(–34) × (8–)10–26 mm, (red-)brown to dark purple-brown (rarely pale pinkish brown).

NOTE. *Serapias occidentalis* C.Venhuis & P.Venhuis from SW Spain consists of populations of plants that are morphologically intermediate between *S. cordigera* and *S. vomeracea* s.s.. It might represent a partly stabilised hybrid complex between these two taxa.

OPPOSITE LEFT TO RIGHT
S. cordigera
France, Haute-Provence 09.05.2005
Sardinia 17.04.2017
Sardinia 17.04.2017

Serapias perez-chiscanoi Acedo, *Anales Jard. Bot. Madrid* 47: 510 (1990). **Perez-Chiscano's tongue-orchid**
DISTRIBUTION. Portugal and SW Spain (Extremadura).
HABITAT. Garigue and damp to marshy grassland on slightly acid soil; sea level to 400 m.
FLOWERING. March to May.
DISTINGUISHING FEATURES. 20–40 cm tall. Inflorescence dense at peak flowering (any bract in the mid-portion of the spike overlapping with the bract immediately above for more than half of its length). Flowers always strongly hypochromic. Lip at the base with two parallel keels; epichile 17–22 × 9–14 mm, pale green.

BOTH IMAGES
Spain, Badajoz 28.04.1996 (1)

Serapias nurrica Corrias, *Boll. Soc. Sarda Sci. Nat.* 21: 397 (1982). **Sardinian tongue-orchid**

DISTRIBUTION. Scattered in the WC Mediterranean region, from Menorca to S Italy.

HABITAT. On acid soil in garigue, moist grassland and open places in maquis; sea level to 1,000 m.

FLOWERING. April to June.

DISTINGUISHING FEATURES. 15–35 cm, or more, tall. Inflorescence dense at peak flowering (any bract in the mid-portion of the spike overlapping with the bract immediately above for more than half of its length). Flowers autogamous (pollinia disintegrating when, or even before, the flower buds open; ovaries swelling while the perianths are still fresh); lip red-brown with conspicuous white margins, with two parallel keels at the base.

BELOW, LEFT TO RIGHT
Sardinia 28.04.1994 (1)
Sicily 25.05.1992 (1)

Serapias neglecta De Not., *Repert. Fl. Ligust.*: 55 (1844). **Scarce tongue-orchid**

DISTRIBUTION. From Sardinia across Corsica to the French and Italian Riviera.

HABITAT. On basic to slightly acid soil in garigue, olive groves, meadows and open oak forest; sea level to 600 m.

FLOWERING. March to May.

DISTINGUISHING FEATURES. 10–35 cm tall. Inflorescence dense at peak flowering (any bract in the mid-portion of the spike overlapping with the bract immediately above for more than half of its length). Flowers allogamous (ovaries usually not swelling while the perianths are still fresh). Lip at the base provided with two parallel keels; hypochile funnel-shaped in the front view; epichile 0.7–0.8 times as broad as the (flattened) hypochile, usually salmon-coloured to ochre-yellow.

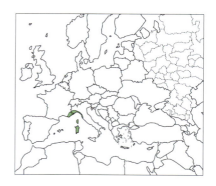

CLOCKWISE, FROM TOP LEFT
France, Provence 12.05.2004
France, Provence 11.05.2003
Italy, Elba 25.03.2009 (21)
France, Provence 13.05.2007
France, Provence 09.05.2005

Serapias parviflora Parl., *Giorn. Sci. Sicilia* 59: 66 (1837). **Small-flowered tongue-orchid**

DISTRIBUTION. Canary Islands, Azores (Sta Maria), W France and the Mediterranean region east to Cyprus. Recorded from SW England in recent years but has since disappeared.

HABITAT. Garigue, scrub, dune slacks, meadows, olive grows and open forest on calcareous to slightly acid soil; sea level to 1,200 m.

FLOWERING. March to early June.

DISTINGUISHING FEATURES. 10–30 cm, or more, tall. Flowers autogamous (pollinia disintegrating when, or even before, the flower buds open; ovaries swelling while the perianths are still fresh). Lip (red-)brown, at the base provided with two parallel keels; epichile 5–10(–13) × 3–5 mm.

NOTE. *Serapias elsae* P.Delforge from C Portugal and SW Spain is somewhat morphologically intermediate between *S. parviflora* and *S. strictiflora*. It might consist of hybrid swarms between these two taxa.

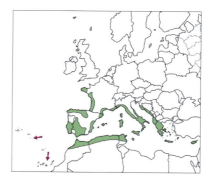

CLOCKWISE, FROM TOP LEFT
Cyprus 23.03.1999
Cyprus 20.03.2009
Cyprus 20.03.2009
Sardinia 14.04.2017
Sardinia 14.04.2017
Sardinia 14.04.2017

Serapias vomeracea (Burm.f.) Briq., *Prodr. Fl. Corse* 1: 378 (1910). **Long-lipped tongue-orchid**
SYNONYM. *Orchis vomeracea* Burm.f.

a. subsp. *vomeracea*

SYNONYM. *Serapias vomeracea* var. *guadarramica* Kreutz

DISTRIBUTION. Widespread in the Mediterranean region from Morocco to the E Aegean Islands, with an eastern outpost in Cyprus; also occurring in adjoining warm-temperate parts of Europe.

HABITAT. Grassland, garigue, olive groves, open woodland and damp meadows on calcareous to slightly acid soil; sea level to 1,500 m.

FLOWERING. Late March (in the south) to early July (in the north).

DISTINGUISHING FEATURES. 15–60 cm tall. Inflorescence lax at peak flowering (any bract in the mid-portion of the spike overlapping with the bract immediately above for less than half of its length, if at all). Flowers allogamous (ovaries usually not swelling while the perianths are still fresh). Lip (red-)brown, at the base provided with two parallel keels; hypochile tunnel- to U-shaped in front view, 17–25 mm broad when flattened; epichile (9–)13–30 × (8–)9–13 mm, conspicuously hairy in its proximal third (or more).

OPPOSITE, LEFT TO RIGHT FROM TOP
subsp. *vomeracea*
Italy, Puglia 24.04.2012 (45)
Spain, Andalucia 16.04.2001 (45)
Cyprus 17.03.2003
Italy, Umbria 23.04.2008 (21)
Cyprus 16.03.2011
Cyprus 16.03.2003
Cyprus 17.03.1995
France, Aveyron 18.05.1987
subsp. *laxiflora*
Greece, Pelion 05.04.2022 (25)
Crete 05.04.2014
Greece, Samos 13.04.1999 (45)

b. *Serapias vomeracea* subsp. *laxiflora* (Soó) Gölz & H.R.Reinhard, *Orchidee (Hamburg)* 28: 114 (1977). **Lax-flowered tongue-orchid**

SYNONYMS. *Serapias bergonii* E.G.Camus; *S. parviflora* subsp. *laxiflora* Soó; *S. cilentana* (Presser) Presser

DISTRIBUTION. From Malta, Sicily and S mainland Italy to W & S Turkey and Cyprus.

HABITAT. Meadows, grassland, olive groves, garigue and open maquis and woodland on calcareous to slightly acid soil; sea level to 1,500 m.

FLOWERING. March to June.

DISTINGUISHING FEATURES. 10–40 cm, or more, tall. Differs from the typical subspecies in the hypochile being 10–15 mm broad (when flattened) and in the epichile being 4–7(–8) mm broad and usually sparsely hairy to subglabrous in its proximal third.

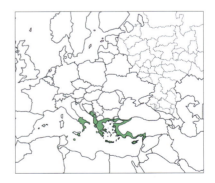

NOTE. *Serapias aphroditae* P.Delforge from W Cyprus and SW Anatolia might deserve taxonomic recognition at infraspecific level. It differs from typical *S. vomeracea* subsp. *laxiflora* in that the epichile is not distinctly longer than the hypochile.

Serapias politisii Renz, reported from Italy, Greece and W Turkey, is commonly recognised as a distinct species. However, available results from morphometric and molecular studies are confusing and seem to indicate that not all populations treated under this name share a common taxonomic identity. We are inclined to think that most colonies assigned to *S. politisii* are hybrid swarms between *S. parviflora* and *S. vomeracea* subsp. *laxiflora*.

Serapias orientalis (Greuter) H.Baumann & Künkele, *Mitt. Arbeitskreis Heimische Orchid. Baden-Württemberg* 20: 636 (1988). **Eastern tongue-orchid**

SYNONYMS. *Serapias vomeracea* subsp. *orientalis* Greuter; *S. ionica* H.Baumann & Künkele; *S. istriaca* Perko; *S. feldwegiana* H.Baumann & Künkele; *S. levantina* H.Baumann & Künkele; *S. apulica* (H.Baumann & Künkele) P.Delforge; *S. orientalis* var. *carica* (H.Baumann & Künkele) P.Delforge; *S. patmia* M.Hirth & H.Spaeth; *S. orientalis* var. *siciliensis* (Bartolo & Pulv.) P.Delforge; *S. orientalis* var. *monantha* (P.Delforge) P.Delforge; *S. orientalis* var. *sennii* (Renz) P.Delforge; *S. orientalis* var. *spaethiae* P.Delforge; *S. levantina* var. *dafnii* (B.Baumann & H.Baumann) P.Delforge

DISTRIBUTION. From Sicily and southern mainland Italy across the E Mediterranean region to W Transcaucasus.

HABITAT. Mainly on basic soil in grassland, meadows, garigue, maquis and olive groves and on abandoned agricultural terraces; sea level to 1,550 m.

FLOWERING. March to June.

DISTINGUISHING FEATURES. 3–50 cm tall. Inflorescence dense at peak flowering (any bract in the mid-portion of the spike overlapping with the bract immediately above for more than half of its length). Flowers allogamous (ovaries usually not swelling while the perianths are still fresh). Lip at the base provided with two parallel keels; hypochile tunnel- to U-shaped in front view; epichile 0.5–0.6(–0.7) times as broad as the (flattened) hypochile, usually dark red-brown.

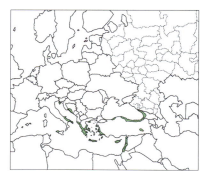

LEFT TO RIGHT FROM TOP
Crete 05.04.2014
Crete 05.04.2014
Crete 25.04.2014
Italy, Puglia 10.04.2016
Italy, Puglia 10.04.2016
Italy, Puglia 10.04.2016

Serapias olbia Verg., *Bull. Soc. Bot. France* 54: 597 (1908). **Hybrid tongue-orchid**

SYNONYM. *Serapias gregaria* Godfery

DISTRIBUTION. Only known with certainty in Corsica and southern Provence (doubtful records exist from Spain and NW Italy).

HABITAT. On acid soil in garigue, meadows, moist dune slacks and open oak forest; sea level to 200 m.

FLOWERING. April and May.

DISTINGUISHING FEATURES. 10–30 cm tall. Lip at the base provided with an elongate callus that is furrowed throughout its length and retuse to emarginate in the front; hypochile slightly protruding from the sepals (in side view); epichile (ovate-)lanceolate, (7–)8–9.5 mm broad, about half as broad as the (flattened) hypochile, distinctly hairy in its basal part, usually red-brown to dark purple-brown.

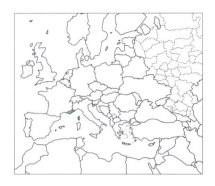

LEFT TO RIGHT FROM ABOVE
France, Provence 05.05.2008
France, Provence 15.05.2004
France, Haute-Provence 15.05.1999
France, Var 23.04.1991 (11)
France, Provence 01.03.1994 (10)

Serapias strictiflora Welw. ex Veiga, *Orch. Port.*: 18 (1886). **Straight-flowered tongue-orchid**

DISTRIBUTION. W Mediterranean region (east to Algeria and possibly Tunisia).

HABITAT. Roadside verges, grassland, meadows, garigue, maquis and open forest on calcareous to acid soil; sea level to 900 m (and possibly higher).

FLOWERING. Late March to May.

DISTINGUISHING FEATURES. 10–40 cm, or more, tall. Lip at the base provided with an elongate callus that is furrowed throughout its length and retuse to emarginate in the front; hypochile in most flowers completely hidden by the sepals (in side view); epichile linear-triangular to linear-lanceolate, 2.5–5(–6) mm broad, about $1/3$ as broad as the (flattened) hypochile, distinctly hairy in its basal part, usually red-brown to dark purple-brown.

NOTE. Certain high-altitude (700–1,400 m) Algerian populations consist of individuals that differ from typical *S. strictiflora* in being more robust, dense-flowered plants with relatively broader epichiles. They have been described as *S. athwaghlisia* Kreutz & Rebbas and might deserve taxonomic recognition.

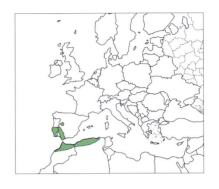

BELOW, LEFT TO RIGHT
Portugal, Algizur 21.04.1996
Portugal, Lisbon 15.04.1979 (1)

Serapias lingua L., *Sp. Pl.*: 950 (1753).
Common tongue-orchid

DISTRIBUTION. W France and widespread in the Mediterranean region from Portugal and Morocco to Greece (and possibly W Turkey).

HABITAT. On basic to acid soil in meadows, garigue and grassland and in open places in woods and maquis; sea level to 1,200 m.

FLOWERING. March to June.

DISTINGUISHING FEATURES. 10–30 cm, or more, tall. Lip at the base provided with an elongate, shiny, basally furrowed callus that is rounded (to truncate) in the front; epichile subglabrous in its basal part, usually pinkish purple to pale ochre-yellow.

BELOW, ALL IMAGES
Sardinia 15.04.2017

OPPOSITE, LEFT TO RIGHT FROM TOP
Sardinia 15.04.2017 (long lip)
Sardinia 15.04.2017 (wide lip)
Sardinia 15.04.2017 (wide lip)
Sardinia 15.04.2017
France, Aveyron 23.05.1996
France, Aveyron 19.05.1996
Crete 07.04.2015
Crete 01.04.2014
Majorca 19.04.2013

30. ANACAMPTIS Rich., *De Orchid. Eur.*: 25 (1817).

SYNONYMS. *Vermeulenia* Á.Löve & D.Löve; *Anteriorchis* E.Klein & Strack; *Herorchis* D.Tyteca & Strack; *Paludorchis* P.Delforge

11 species in Europe and the E Mediterranean, C and SW Asia and N Africa.

Until recently, *Anacamptis* was treated as a monotypic genus based upon *A. pyramidalis*. DNA-based studies have shown that a number of species previously referred to *Orchis* are more closely related to *Anacamptis* than to *Orchis*. Although the individual species are easy to recognise, it is difficult to distinguish *Anacamptis* from *Orchis* at the generic level. However, the combination of at least three leaves usually being placed on the stem above the basal rosette (if any) and a lip that consistently lacks tufts of tiny, coloured hairs separates *Anacamptis* from *Orchis* fairly well. Identification key on p. 433.

Anacamptis laxiflora (Lam.) R.M.Bateman, Pridgeon & M.W.Chase, *Lindleyana* 12: 120 (1997).
Lax-flowered orchid

SYNONYMS. *Orchis laxiflora* Lam.; *O. palustris* subsp. *laxiflora* (Lam.) Batt.; *Paludorchis laxiflora* (Lam.) P.Delforge

a. subsp. *laxiflora*

DISTRIBUTION. Channel Islands to WC Europe, Mediterranean Europe across to Cyprus, Turkey and the Crimea.

HABITAT. Coastal marshes, seepages and flushes on hillsides and wet meadows, in calcareous to slightly acidic soils; sea level to 1,600 m.

FLOWERING. Late April to June.

DISTINGUISHING FEATURES. 30–60 cm tall. Stem usually violet-tinged. Leaves arranged along the stem, keeled, broadest below the middle. Lateral sepals erect or reflexed. Lip shallowly 3-lobed, usually devoid of markings, its basal to central part usually white and making a strong contrast to the reflexed, deep violet-purple (rarely paler) side lobes; spur weakly upcurved.

LEFT TO RIGHT FROM TOP
subsp. *laxiflora*
Greece, Peloponnese 06.04.2019 (25)
France, Aveyron 18.05.2010
Cyprus 16.03.2012
France, Aveyron 16.05.2001
France, Aveyron 20.05.2009
France, Aveyron 20.05.2009
France, Aveyron 20.05.1989
France, Var 13.05.1996
France, Var 13.05.1996
France, Var 13.05.1996

b. *Anacamptis laxiflora* subsp. *dielsiana* (Soó) H.Kretzschmar, Eccarius & H.Dietr., *Orchid Gen. Anacamptis Orchis Neotinea*: 98 (2007).

SYNONYMS. *Orchis laxiflora* subsp. *dielsiana* Soó; *O. pseudolaxiflora* Czerniak.; *O. palustris* subsp. *pseudolaxiflora* (Czerniak.) H.Baumann & R.Lorenz; *Paludorchis pseudolaxiflora* (Czerniak.) P.Delforge

DISTRIBUTION. From E Turkey across to Afghanistan and Arabia.

HABITAT. Marshes and wet grassland; 700–2,300 m.

FLOWERING. May and June.

DISTINGUISHING FEATURES. 30–105 cm tall. Similar to the typical subspecies but differs in having pink flowers with a lip with spreading side lobes and a straight spur.

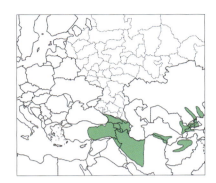

c *Anacamptis laxiflora* subsp. *dinsmorei* (Schltr.) H.Kretzschmar, Eccarius & H.Dietr., *Orchid Gen. Anacamptis Orchis Neotinea*: 101 (2007).

SYNONYMS. *Orchis laxiflora* var. *dinsmorei* Schltr.; *O. elegans* var. *dinsmorei* (Schltr.) H.I.Schäf.; *O. dinsmorei* (Schltr.) H.Baumann & Dafni; *O. laxiflora* subsp. *dinsmorei* (Schltr.) Kreutz; *Paludorchis dinsmorei* (Schltr.) P.Delforge

DISTRIBUTION. E Mediterranean from SC Turkey to Israel.

HABITAT. Damp meadows, flushes and marshes; up to 1,000 m.

FLOWERING. March and April.

DISTINGUISHING FEATURES. 45–70 cm tall. Similar to the typical subspecies but differs in having purple flowers and a lip with not greatly recurved side lobes. Besides, the inflorescence is generally denser and the flowers generally smaller, almost consistently with markings on the lip and a more strongly upcurved spur.

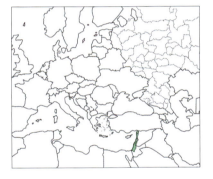

LEFT TO RIGHT FROM TOP
subsp. *dielsiana*
Turkey, Adiyaman 14.05.1988 (56)
Turkey, Adiyaman 21.05.1995 (56)
Turkey, Adiyaman 14.05.1988 (1)
Turkey, Adiyaman 14.05.1988 (1)
subsp. *dinsmorei*
Turkey, Gazientep 23.05.1997 (56)
Israel, Hagoscherim 28.04.1992 (56)
Turkey, Gazientep 23.05.1997 (56)
Israel, Hagoscherim 28.04.1992 (56)
Turkey, Gazientep 23.05.1997 (56)

Anacamptis palustris (Jacq.) R.M.Bateman, Pridgeon & M.W.Chase, *Lindleyana* 12: 120 (1997).
SYNONYMS. *Orchis palustris* Jacq.; *O. laxiflora* var. *palustris* (Jacq.) Mérat; *O. laxiflora* subsp. *palustris* (Jacq.) Bonnier & Layens; *Paludorchis palustris* (Jacq.) P.Delforge

a. subsp. *palustris*

DISTRIBUTION. N Tunisia and most of Europe north to the English Channel and Gotland.
HABITAT. In moist soil and marshes, usually on basic soils; up to 1,800 m.
FLOWERING. April to July.
DISTINGUISHING FEATURES. 20–60 cm, or rarely more, tall. Leaves arranged along the stem, keeled, broadest below the middle. Lateral sepals erect or reflexed. Lip usually less than 14 mm long, distinctly 3-lobed, marked with dots and dashes in its median part which is pale but does not contrast strongly with the rose-pink, more or less spreading side lobes; mid-lobe usually longer than the side lobes; spur straight to weakly downcurved (very rarely weakly upcurved), slightly longer than the lip blade, usually horizontal.

b. *Anacamptis palustris* subsp. *elegans* (Heuff.) R.M.Bateman, Pridgeon & M.W.Chase, *Lindleyana* 12: 120 (1997).

SYNONYMS. *Orchis elegans* Heuff.; *O. palustris* var. *elegans* (Heuff.) Nyman; *Paludorchis palustris* var. *elegans* (Heuff.) P.Delforge
DISTRIBUTION. Balkans north to Hungary, Slovakia and S Russia, and from Greece across Turkey and Cyprus to Iran and NE Saudi Arabia.
HABITAT. In moist soil and marshes, usually on basic soils; up to 1,800 m.
FLOWERING. April to June.
DISTINGUISHING FEATURES. 50–80 cm tall. Very similar to the typical subspecies but differs in having deep magenta to purple flowers, an obscurely to shallowly 3-lobed lip in which the mid-lobe is usually shorter than the side lobes, and a suberect spur.

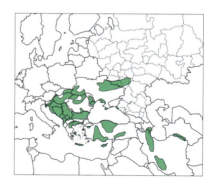

LEFT TO RIGHT FROM TOP
subsp. *palustris*
Switzerland, Bern 06.06.1996
Switzerland, Bern 03.06.2002
Switzerland, Bern 03.06.2002
Switzerland, Bern 03.06.2002
Switzerland, Bern 06.06.1996
subsp. *elegans*
Turkey, Antalya 23.05.2013 (40)
Turkey, Antalya 23.05.2013 (40)
Greece, Lesbos 26.05.2006 (14)

ANACAMPTIS

c. *Anacamptis palustris* subsp. *robusta*
(T.Stephenson) R.M.Bateman, Pridgeon & M.W.Chase, *Lindleyana* 12: 120 (1997).
SYNONYMS. *Orchis palustris* var. *robusta* T.Stephenson; *O. robusta* (T.Stephenson) Gölz & H.R.Reinhard; *Paludorchis robusta* (T.Stephenson) P.Delforge
DISTRIBUTION. Majorca and N Algeria.
HABITAT. In marshes; up to 200 m.
FLOWERING. March to early May.
DISTINGUISHING FEATURES. Robust, up to 90 cm tall with leaves up to 30 cm long. Similar to the typical subspecies but differs in having a lip that is usually more than 14 mm long and a somewhat downwards pointing spur that is shorter than the lip blade.

LEFT TO RIGHT
subsp. *robusta*
Majorca 15.04.2013
Majorca 15.04.2013
Majorca 15.04.2013

Anacamptis collina (Banks & Sol. ex Russell)
R.M.Bateman, Pridgeon & M.W.Chase, *Lindleyana* 12: 120 (1997). **Saccate-lipped orchid**
SYNONYMS. *Orchis collina* Banks & Sol. ex Russell; *O. saccata* Ten.; *O. sparsiflora* Spruner ex Rchb.f.; *O. chlorotica* Woronow; *O. fedtschenkoi* Czerniak.; *O. leucoglossa* O.Schwarz; *O. dulukae* Hautz.; *Vermeulenia collina* (Banks & Sol. ex Russell) P.Delforge
DISTRIBUTION. Mediterranean (with major gaps) to S Turkmenistan.
HABITAT. In full sun in stony grassland, openings in scrub and woodlands on calcareous soils; up to 1,300 m.
FLOWERING. January to April.
DISTINGUISHING FEATURES. 10–40 cm tall, stout. Lower leaves assembled in a basal rosette. Sepals and petals green, often flushed purple or brown. Lip entire, white to dark purple with pale base, similar to that of a small *A. papilionacea* but lacking coloured veins; spur very thick, 5–7 mm long, down-pointing.

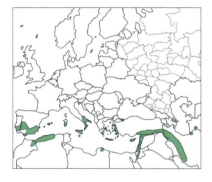

LEFT TO RIGHT
Crete 10.04.2014
Crete 10.04.2014
Cyprus 12.03.2000
Cyprus 18.03.2009
Cyprus 10.03.2002
Crete 02.04.2014
Crete 02.04.2014
Crete 02.04.2014

ANACAMPTIS

Anacamptis pyramidalis (L.) Rich., *De Orchid. Eur.*: 33 (1817). **Pyramidal orchid**

SYNONYMS. *Orchis pyramidalis* L.; *Anacamptis pyramidalis* var. *brachystachys* (d'Urv.) Boiss.; *A. pyramidalis* var. *tanayensis* Chenevard; *A. pyramidalis* var. *nivea* P.Delforge; *A. pyramidalis* var. *cerigensis* P.Delforge; *A. pyramidalis* var. *serotina* (H.Presser) Kreutz

DISTRIBUTION. Europe north to Scotland, Gotland and Estonia; Mediterranean region across to N Iran.

HABITAT. Grassland, open places in scrub and woodland, waste places, often in calcareous soils in full sun to light shade; up to 2,000 m.

FLOWERING. April to late July.

DISTINGUISHING FEATURES. Inflorescence at first pyramidal, later ovoid, of uniformly rich purple to pink or rarely white small flowers with a 3-lobed

RIGHT
A. pyramidalis var. *tanayensis*
Switzerland, Aargau 30.06.1995

THIS PAGE, LEFT TO RIGHT
France, Alsace 01.06.2017
Italy, Tuscany 14.05.2012
Switzerland, Aargau 29.05.2017
France, Alsace 03.06.2010
France, Aveyron 18.05.1992
France, Aveyron 20.05.2004
France, Aveyron 20.05.2004
France, Alsace 30.06.2009 (25)
France, Aveyron 20.05.2004
France, Aveyron 18.05.1990

(very rarely entire) lip bearing a short, raised callus on each side of the opening of the slender spur.

NOTE. Variation in ploidy level, molecular markers, flower colour, morphology, anatomy, flowering time and habitat requirements has led to description of taxa such *A. pyramidalis* var. *tanayensis* from Switzerland, *A. urvilleana* Sommier & Caruana from Malta and *A. iberico* Doro from mainland Italy. However, as the patterns of variation appear to be highly complex, we currently abstain from formally recognizing any proposed (sub)division of *A. pyramidalis* s.l.. A comprehensive biosystematic study conducted on a broad geographic scale is long overdue.

Anacamptis sancta (L.) R.M.Bateman, Pridgeon & M.W.Chase, *Lindleyana* 12: 120 (1997). **Holy orchid**

SYNONYMS. *Orchis sancta* L.; *O. coriophora* var. *sancta* (L.) Rchb.f.; *O. coriophora* subsp. *sancta* (L.) Hayek.; *Anteriorchis sancta* (L.) E.Klein & Strack

DISTRIBUTION. E Mediterranean from Aegean islands, Crete east to coastal W and S Turkey, Lebanon, Palestine and Israel.

HABITAT. In open stony places in scrub and waste ground.

FLOWERING. Late April to early July.

DISTINGUISHING FEATURES. 20–50 cm tall. Similar to *A. coriophora* subsp. *fragrans* but flowering later with larger more uniformly pink flowers without spotting on the lip. Besides, the dorsal sepal is long-acuminate to caudate, the lip is not abruptly downcurved at the base, and the spur mouth has a pair of characteristic triangular processes.

BELOW, LEFT TO RIGHT
Patmos 17.04.1995 (1)
Rhodes 24.04.1966 (10)
Rhodes 24.04.1966 (10)
Lesbos 25.05.2006 (14)

Anacamptis coriophora (L.) R.M.Bateman, Pridgeon & M.W.Chase, *Lindleyana* 12: 120 (1997).
Bug orchid

SYNONYMS. *Orchis coriophora* L.; *Anteriorchis coriophora* (L.) E.Klein & Strack

a. subsp. *coriophora*

SYNONYMS. *Anteriorchis coriophora* var. *carpetana* (Willk.) P.Delforge; *A. coriophora* var. *martrinii* (Timb.-Lagr.) P.Delforge

DISTRIBUTION. C Europe (north to Germany and Belarus), Mediterranean to W Asia. In subtropical regions, it is largely restricted to mountains.

HABITAT. In full sun in slightly acid to slightly basic soils in grassland, meadows, stony places and marshy moors; sea level to 2,500 m.

FLOWERING. April to July.

DISTINGUISHING FEATURES. 15–50 cm tall. Lower leaves assembled in a basal rosette. Flowers unpleasantly smelling of bedbugs, opening from the bottom of the inflorescence upwards. Sepals and petals connivent to form a hood over the column; dorsal sepal acute to somewhat acuminate. Lip abruptly downcurved at the base, broader than long, ruby to blackish-ruby (rarely green) with a pale, usually purple-spotted base, distinctly 3-lobed; side lobes spreading, their hind edges perpendicular to the lip base; spur downcurved, 4–8 mm long, spur-mouth devoid of triangular processes.

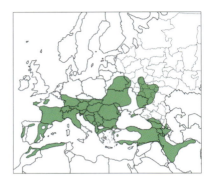

OPPOSITE, LEFT TO RIGHT
subsp. *coriophora*
France, Languedoc-Roussillon 26.05.2003
Turkey, Antalya 22.05.2013 (40)
France, Languedoc-Roussillon 26.05.1996
France, Languedoc-Roussillon 20.05.1999
France, Les Maze 18.05.2000
subsp. *fragrans*
Italy, Puglia 14.04.2016
Italy, Puglia 14.04.2016
Majorca 20.04.2013
Italy, Puglia 14.04.2016

b. *Anacamptis coriophora* subsp. *fragrans*
(Pollini) R.M.Bateman, Pridgeon & M.W.Chase, *Lindleyana* 12 (3): 120 (1997). **Fragrant bug orchid**

SYNONYMS. *Orchis fragrans* Pollini; *O. coriophora* var. *fragrans* (Pollini) Boiss; *O. coriophora* subsp. *fragrans* (Pollini) K.Richt; *Anteriorchis fragrans* (Pollini) Szlach.

DISTRIBUTION. Widespread in the Mediterranean region from where it extends to Brittany and Iraq. Grows in warmer climates than the typical subspecies.

HABITAT. Stony open places in dry calcareous soils in poor grassland, scrub and open woodland; up to 1,100 m.

FLOWERING. April to July.

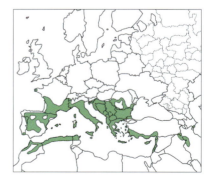

ANACAMPTIS

DISTINGUISHING FEATURES. 10–50 cm tall. Differs from the typical subspecies in having flowers that smell pleasantly of vanilla and a lip at least as long as broad. Furthermore, the dorsal sepal is more acuminate and the side lobes of the lip are recurved, with their hind edges at obtuse angles to the lip base.

TOP TO BOTTOM, LEFT TO RIGHT
A. coriophora subsp. *fragrans*
Majorca 20.04.2013
Cyprus 17.03.2007
France, Var 20.05.1988
France, Aveyron 20.05.2009
Greece, Smolikas 02.07.1994 (14)
Turkey, Antalya 22.05.2013 (40)
Majorca 20.04.2013
Crete 22.05.2001 (14)

Anacamptis cyrenaica (E.A.Durand & Barratte) H.Kretzschmar, Eccarius & H.Dietr., *Orchid Gen. Anacamptis Orchis Neotinea*: 178 (2007).
Libyan butterfly orchid
SYNONYMS. *Orchis cyrenaica* E.A.Durand & Barratte; *O. papilionacea* var. *cyrenaica* (E.A.Durand & Barratte) P.Delforge; *Vermeulenia cyrenaica* (E.A.Durand & Barratte) P.Delforge
DISTRIBUTION. Endemic to NE Libya.
HABITAT. Open scrub and stony places; up to 300 m.
FLOWERING. Late February to mid-April.
DISTINGUISHING FEATURES. Plant 20–35 cm tall. Lower leaves assembled in a rosette. Lateral sepals somewhat spreading, with conspicuous veins. Lip subentire with spreading to recurved sides, broadest above the middle, gradually tapering towards base, pink with pale to white base, marked with a pair of dark purple spots at the base and usually with a row of minor markings along the mid-line; spur downcurved, slender, tapering towards the apex, approximately as long as the ovary.

BELOW, CLOCKWISE FROM LEFT
Libya 12.03.2002 (2)
Libya 12.03.2002 (2)
Libya 29.03.2000 (1)

Anacamptis papilionacea (L.) R.M.Bateman, Pridgeon & M.W.Chase, *Lindleyana* 12: 120 (1997).
Pink butterfly orchid

SYNONYMS. *Orchis papilionacea* L.; *Vermeulenia papilionacea* Á.Löve & D.Löve

a. subsp. *papilionacea*

SYNONYMS. *Orchis rubra* Jacq.; *O. papilionacea* subsp. *heroica* E.D.Clarke auct.; *O. papilionacea* var. *messenica* Renz; *O. papilionacea* subsp. *alibertis* G. Kretzschmar & H.Kretzschmar; *Vermeulenia bruhnsiana* (Gruner) Szlach.; *O. papilionacea* subsp. *balcanica* H.Baumann & R.Lorenz; *Anacamptis papilionacea* subsp. *schirwanica* (Woronow) H.Kretzschmar, Eccarius & H.Dietr.; *Vermeulenia papilionacea* var. *alibertis* (G. Kretzschmar & H.Kretzschmar) P.Delforge; *V. papilionacea* var. *messenica* (Renz) P.Delforge; *Anacamptis papilionacea* subsp. *thaliae* Kreutz, J.Essink & L.Essink; *Vermeulenia papilionacea* var. *aegaea* P.Delforge; *Anacamptis papilionacea* subsp. *rubra* (Jacq.) Pérez-Chisc. & J.P.Prieto

DISTRIBUTION. Mediterranean SE France, Italy, Corsica, Sardinia and Sicily, the Balkans and Greece across to Turkey and Azerbaijan.

HABITAT. In full sun and light shade in dry, alkaline soils in stony grassland, garigue and open woodland; up to 1,800 m.

FLOWERING. February to May.

DISTINGUISHING FEATURES. 15–40 cm tall. Lower leaves assembled in a basal rosette. Inflorescence hemispherical to conical (rarely cylindrical), flowers opening from the base upwards, pink to purple-red with white to pink lip. Sepals and petals with prominent veins, connivent or sepals somewhat spreading. Lip uniformly coloured or marked with radiating purple lines, subspathulate to subcircular, crenate, 12–25 × 11–25 mm, at least as long as wide, with moderately to strongly incurved sides, cuneate to subtruncate at the base; spur downcurved, tapering towards the apex.

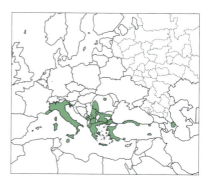

OPPOSITE PAGE, LEFT TO RIGHT
subsp. *papilionacea*
France, Linas 19.05.2006
France, Provence 18.05.2003
France, Provence 15.05.2007
Italy, Siena 18.04.2000
Italy, Siena 27.04.2009
Italy, Siena 25.04.2008
Sardinia 17.04.2017
SW Turkey 05.1988 (22)
SW Turkey 05.1988 (22)

RIGHT
Italy, Puglia 14.04.2016

b. *Anacamptis papilionacea* subsp. *expansa*
(Ten.) Amardeilh & Dusak, *L'Orchidophile* 165: 104 (2005).

SYNONYMS. *Orchis expansa* Ten.; *O. papilionacea* var. *grandiflora* Boiss.; *Vermeulenia papilionacea* var. *grandiflora* (Boiss.) Szlach.; *V. papilionacea* var. *vexillifera* (Terraciano) P.Delforge

DISTRIBUTION. W Mediterranean east to Sicily and mainland S Italy.

HABITAT. In full sun and light shade in dry, alkaline soils in stony grassland, garigue and open woodland; up to 1,800 m.

FLOWERING. February to May.

DISTINGUISHING FEATURES. A robust plant, 15–40 cm tall. Similar to the typical subspecies but differs in having a lip that is wider than long, with spreading to weakly incurved sides, and consistently subtruncate at the base. Besides, the lip is generally larger (15–26 × 16–30 mm) and virtually always marked with radiating lines.

c. *Anacamptis papilionacea* subsp. *palaestina*
(H. Baumann & R.Lorenz) H.Kretschmar, Eccarius & H.Dietr., *Orch. Gen. Anacamptis, Orchis, Neotinea*: 164 (2007).

SYNONYMS. *Orchis papilionacea* subsp. *palaestina* H.Baumann & R.Lorenz; *Vermeulenia papilionacea* var. *palaestina* (H.Baumann & R.Lorenz) P.Delforge

DISTRIBUTION. S Turkey to Israel and Palestine; possibly also in Cyprus.

HABITAT. In full sun and light shade in dry, alkaline soils in open pine woodland; up to 1,000 m.

FLOWERING. February to May.

DISTINGUISHING FEATURES. 15–40 cm tall. Similar to the typical subspecies but differs in having a generally smaller lip (usually less than 12 mm long) with spreading sides and being marked partly with spots, rather than being either devoid of markings or marked with radiating lines only. Besides, the inflorescence is usually cylindrical and with more flowers.

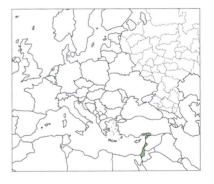

OPPOSITE PAGE, LEFT TO RIGHT
subsp. *expansa*
France, Provence 17.05.2009
Sardinia 02.04.2016
Sicily 17.05.2009
Sardinia 23.04.1997 (45)
France, Aveyron 22.05.2009
France, Aveyron 22.05.2009

subsp. *palaestina*
Israel, Mt Carmel 28.02.1989 (22)
Israel 05.1993 (22)
Israel 05.1993 (22)

Anacamptis boryi (Rchb.f.) R.M.Bateman, Pridgeon & M.W.Chase, *Lindleyana* 12: 120 (1997).
Bory's orchid
SYNONYMS. *Orchis boryi* Rchb.f.; *O. quadripunctata* var. *boryi* (Rchb.f.) Nyman; *Herorchis boryi* (Rchb.f.) D.Tyteca & E.Klein
DISTRIBUTION. SC & S mainland Greece, Aegean Islands and Crete.
HABITAT. In grassland and open scrub in full sun and light shade; up to 1,300 m.
FLOWERING. April and May.
DISTINGUISHING FEATURES. Plants 10–35 cm tall. Lower leaves assembled in a basal rosette. Flowers dark purple, rarely, pink or white, opening from the top of inflorescence downwards. Sepals and petals connivent to form a loose hood over the column. Lip fan-shaped with spreading sides, broadest above the middle, entire to moderately 3-lobed, usually marked with a few spots in its basal to central part; spur 12–18 mm long, weakly downcurved.

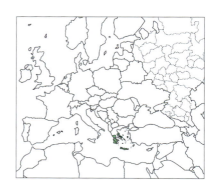

LEFT TO RIGHT, FROM TOP
Crete 07.04.2014
Crete 07.04.2013
Crete 05.04.2014
Crete 07.04.2014
Crete 04.04.2014
Crete 04.04 2014

Anacamptis israelitica (H.Baumann & Dafni) R.M.Bateman, Pridgeon & M.W.Chase, *Lindleyana* 12: 120 (1997). **Israeli green-winged orchid**
SYNONYMS. *Orchis israelitica* H.Baumann & Dafni.; *Herorchis israelitica* (H.Baumann & Dafni) D.Tyteca & E.Klein
DISTRIBUTION. Endemic to N Israel.
HABITAT. Damp meadows, scrub and open oak and pine woods on calcareous soils in full sun; 400 to 800 m.
FLOWERING. January to early April.
DISTINGUISHING FEATURES. 10–35 cm tall. Lower leaves assembled in a basal rosette. Flowers with a light-coloured hood, opening from the top of the inflorescence downwards; sepals and petals connivent to form a loose hood over the column; lip whitish-rose to pale pink with a white centre and 3–6 dark purple spots at the base; spur 11–13 mm long.

LEFT TO RIGHT
Israel, Jodefad 06.03.1989 (1)
Israel, Sasa 16.03.1991 (10)
Israel, Coli 03.1997 (21)

Anacamptis morio (L.) R.M.Bateman, Pridgeon & M.W.Chase, *Lindleyana* 12: 120 (1997).
Green-winged orchid
SYNONYMS. *Orchis morio* L.; *Herorchis morio* (L.) D.Tyteca & E.Klein

a. subsp. *morio*
DISTRIBUTION. Europe north to Scotland, S Norway and Estonia, SE to the northern Balkans and east to the Ukraine.
HABITAT. Grassland, meadows, alpine pastures and woodland margins; up to 2,000 m.
FLOWERING. March to early June.
DISTINGUISHING FEATURES. 10–30 cm, or more, tall. Lower leaves assembled in a basal rosette. Flowers in a dense spike that occupies the upper ¼ or more of the stem, opening from the base of the inflorescence upwards. Sepals with green veins, connivent with petals to form a hood over the column. Lip more than 1.5 times as broad as long, shallowly 3-lobed, purple to rose (rarely white), the median part normally spotted and often with pale ground colour but not making a sharp contrast to the side lobes; mid-lobe not distinctly shorter than the spreading to somewhat recurved side lobes; spur upcurved, not tapering towards the apex, at least 1.4 times as long as the lip blade, usually of same colour as the side lobes.

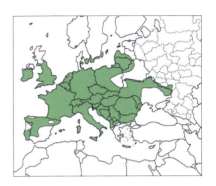

FROM OPPOSITE TOP, LEFT TO RIGHT
subsp. *morio*
Italy, Puglia 10.04.2016
Italy, Puglia 10.04.2016
France, Aveyron 15.05.2001
Italy, Puglia 13.04.2016
France, Aveyron 12.05.1994
Germany, Black Forest 26.05.1992
Germany, Black Forest 12.05.1999
France, Aveyron 22.05.1998
Italy, Puglia 15.04.2016
France Aveyron 26.05.1992
  (with *Eucera longicornis*)
France, Aveyron 12.05.2001 (14)
France, Aveyron 12.05.2001 (15)

b. *Anacamptis morio* subsp. *caucasica* (K.Koch) H.Kretzschmar, Eccarius & H.Dietr., *Orchid Gen. Anacamptis Orchis Neotinea*: 125 (2007).

SYNONYMS. *Orchis morio* var. *caucasica* K.Koch; *O. picta* var. *caucasica* (K.Koch) P.Delforge; *O. albanica* Gölz & H.R.Reinhard; *O. picta* var. *albanica* (Gölz & H.R.Reinhard) P.Delforge; *Anacamptis morio* subsp. *albanica* (Gölz & H.R.Reinhard) Kreutz; *Herorchis picta* var. *albanica* (Gölz & H.R.Reinhard) P.Delforge; *H. picta* var. *caucasica* (K.Koch) P.Delforge; *H. picta* var. *skorpili* (Velen.) P.Delforge

DISTRIBUTION. Balkan Peninsula to the Ukraine and N Iran.

HABITAT. Seasonally wet meadows, open pine woods; up to 1,500 m.

FLOWERING. April to June.

DISTINGUISHING FEATURES. 15–30 cm tall. Similar to the typical subspecies but differs in that the lip is less than 1.5 times as broad as long. Besides, it is a spindlier plant with a lip mid-lobe that is consistently longer than the side lobes.

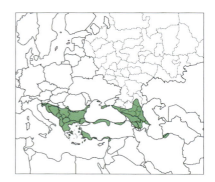

BELOW, LEFT TO RIGHT
subsp. *caucasica*
Greece 12.05.2003 (14)
Greece 12.05.2003 (14)
Greece 02.05.2005 (14)

c. *Anacamptis morio* subsp. *champagneuxii*

(Barnéoud) H.Kretzschmar, Eccarius & H.Dietr., *Orchid Gen. Anacamptis Orchis Neotinea*: 130 (2007).

SYNONYMS. *Orchis champagneuxii* Barnéoud; *O. picta* var. *champagneuxii* (Barnéoud) Nyman; *O. morio* var. *champagneuxii* (Barnéoud) J.A.Guim.; *O. morio* subsp. *champagneuxii* (Barnéoud) E.G.Camus; *Anacamptis champagneuxii* (Barnéoud) R.M.Bateman, Pridgeon & M.W.Chase; *Herorchis champagneuxii* (Barnéoud) D.Tyteca & E.Klein

DISTRIBUTION. S France to NW Africa.

HABITAT. Wet areas in meadows; up to 1,500 m.

FLOWERING. March to early June.

DISTINGUISHING FEATURES. 10–30 cm tall, often in groups. Similar to the typical subspecies but differs in that the lip is provided with no or only obscure markings and has a conspicuous white median zone that contrasts strongly with the reflexed (rather than spreading to recurved) side lobes. Besides, it is a plant of spindly stature. Also similar to subsp. *longicornu*, but in this taxon the spur is paler than the side lobes of the often distinctly spotted lip.

CLOCKWISE FROM RIGHT
subsp. *champagneuxii*
France, Aveyron 18.05.2009
France, Aveyron 20.05.1998
France, Aveyron 17.05.2006
Spain, Andalucia 18.04.2001 (45)
Spain, Cuenca 24.05.2002 (14)

d. *Anacamptis morio* subsp. *longicornu* (Poir.) H.Kretzschmar, Eccarius & H.Dietr., *Orchid Gen. Anacamptis Orchis Neotinea*: 134 (2007).

SYNONYMS. *Orchis longicornu* Poir.; *O. morio* var. *longicornu* (Poir.) Knoche; *Anacamptis longicornu* (Poir.) R.M.Bateman, Pridgeon & M.W.Chase; *Herorchis longicornu* (Poir.) D.Tyteca & E.Klein

DISTRIBUTION. W Mediterranean, from N Algeria and N Tunisia across Malta, Sicily, Sardinia, the Balearic Islands and Corsica to southern mainland France.

HABITAT. Meadows, open stony places in scrub, and montane pastures; up to 2,000 m.

FLOWERING. February to May.

DISTINGUISHING FEATURES. 10–35 cm tall. Similar to the typical subspecies but differs in that the lip has a conspicuous white median zone that contrasts strongly to the reflexed, usually very dark purple side lobes and in that the spur is paler than the side lobes. Also similar to subsp. *champagneuxii*, but this taxon has a lip with magenta to purple pink side lobes and no or only obscure markings.

FROM OPPOSITE TOP LEFT TO RIGHT

subsp. *longicornu*
Italy, Sardinia 13.04.2018
Italy, Sardinia 12.04.2018
Italy, Sardinia 13.04.2017
Italy, Sicily 09.04.22
Italy, Sardinia 10.04.2017
Italy, Sardinia 14.04.2017
Italy, Sardinia 12.04.2017
Italy, Sardinia 12.04.2017
Italy, Sardinia 13.04.2017
Italy, Sardinia 13.04.2017
Italy, Sicily 09.04.2022

e. *Anacamptis morio* subsp. *picta* (Loisel.)
Jacquet & Scappat., *Répartit. Orchid. Sauvages France* (Ed. 3) 3: 7 (2003).

SYNONYMS. *Orchis picta* Loisel.; *O. longicornu* var. *picta* (Loisel.) Lindl.; *Anacamptis picta* (Loisel.) R.M.Bateman; *Herorchis picta* (Loisel.) P.Delforge

DISTRIBUTION. Iberian Peninsula, mainland S France and Corsica.

HABITAT. Meadows, montane pastures and woodland margins; up to 1,400 m.

FLOWERING. March to May.

DISTINGUISHING FEATURES. 10–30 cm tall. Similar to the typical subspecies but differs in having a lax spike that occupies less than 1/4 of the stem and a lip in which the mid-lobe is shorter than the side lobes.

f. *Anacamptis morio* subsp. *syriaca* (E.G.Camus)
H.Kretzschmar, Eccarius & H.Dietr., *Orchid Gen. Anacamptis Orchis Neotinea*: 142 (2007).

SYNONYMS. *Orchis morio* subsp. *syriaca* E.G.Camus; *O. picta* subsp. *libani* Renz; *O. syriaca* Boiss. ex H.Baumann & Künkele; *Anacamptis syriaca* (Boiss. ex H.Baumann & Künkele) R.M.Bateman, Pridgeon & M.W.Chase; *Herorchis syriaca* (Boiss. ex H.Baumann & Künkele) D.Tyteca & E.Klein

DISTRIBUTION. Cyprus, S Turkey to Israel.

HABITAT. Calcareous scrub, open woodland and olive groves; up to 1,300 m.

FLOWERING. February to April.

DISTINGUISHING FEATURES. 10–30 cm tall. Similar to the typical subspecies and subsp. *caucasica* but differs from both in having a white to marginally pale pink lip that is consistently devoid of markings and with a spur that is usually more strongly coloured than the lip blade.

BELOW, LEFT TO RIGHT
subsp. *syriaca*
Turkey, Antalya 19.05.2013 (40)
Turkey, Antalya 19.05.2013 (40) (abnormal form)

OPPOSITE PAGE, TOP LEFT TO RIGHT
subsp. *picta*
France, Aveyron 22.05.2003
France, Aveyron 22.05.2003
France, Aveyron 22.05.2008
Portugal, Algarve 24.03.1999 (14)
France, Aveyron 25.05.1999

subsp. *syriaca*
Cyprus 17.03.1995
Cyprus 15.03.2008
Cyprus 12.03.2011
Turkey, Antalya 19.05.2013 (40)
Cyprus 17.03.1995
Cyprus 16.03.2001
Cyprus 16.03.2009

NATURAL HYBRIDS

We present here a selection of natural hybrids found in Europe and the Mediterranean region. The hybrids covered here fall into two categories:

Infrageneric hybrids are listed here using formulae, comprising the putative parental names separated by a multiplication sign, e.g. *Ophrys umbilicata* × *O. scolopax* and *Orchis militaris* × *O. purpurea*. For the following hybrids, the published name where validated, such as *Dactylorhiza* × *legrandiana* and *Ophrys* × *hybrida*, is given in brackets after the formula.

Intergeneric hybrids are also listed using a formula, e.g. *Anacamptis laxiflora* × *Serapias vomeracea*, with their relevant valid hybrid generic name, e.g. × *Gymnanacamptis* (*Gymnadenia* × *Anacamptis*) or × *Dactylodenia* (*Dactylorhiza* × *Gymnadenia*), followed by a distinctive epithet in brackets after each formula.

The names for the hybrids are in general taken from the *Plants of the World Online* (Govaerts et al., 2023). We also use some names which are categorised as "unplaced" there. For these, we have taken the name from the *List of Hybrids of European Orchids* (Günther, 2023 – www.guenther-blaich.de/hybnaminen.htm).

INFRAGENERIC HYBRIDS

Anacamptis laxiflora × *A. morio*

(*Anacamptis* × *alata* (Fleury) H. Kretzschmar, Eccarius & H. Dietr., *Orchid Gen.* Anacamptis Orchis Neotinea 429 (2007)).

NOTE. Can be a common hybrid. All subspecies of *A. morio* hybridise with those of *A. laxiflora* and backcrosses occur so that hybrids can be difficult to identify.

Anacamptis morio × *A. papilionacea*

(*Anacamptis* × *nicodemi* (Cirillo ex Ten.) B.Bock, *Bull. Soc. Bot. Centre-Ouest* 42: 267 (2012)).

NOTE. Known from Italy and Croatia but possibly elsewhere in the Mediterranean.

Anacamptis morio × *A. pyramidalis*

(*Anacamptis* × *laniccae* Kretzschmar, Eccarius & H.Dietr., *Orchid Gen.* Anacamptis Orchis Neotinea 430 (2007)).

NOTE. A rare hybrid with the inflorescence shape of the latter parent and the lip shape of the former.

Dactylorhiza incarnata × *D. maculata*

(*Dactylorhiza* × *carnea* (E.G.Camus ex Fourny) Soó, *Nom. Nov. Gen.* Dactylorhiza: 9 (1962)).

LEFT TO RIGHT, FROM TOP
Anacamptis laxiflora × *A. morio*
France, Aveyron
France, Aveyron

Anacamptis morio × *A. papilionacea*
Italy, Tuscany
France, Var
Italy, Puglia
France, Var

Anacamptis morio × *A. pyramidalis*
France, Aveyron
France, Aveyron

Dactylorhiza incarnata × *D. maculata*
Switzerland, Tessin
France, Aveyron

Dactylorhiza incarnata ×
D. majalis subsp. lapponica

(*Dactylorhiza* ×*weissenbachiana* Perko, *Carinthia* II 184/104, 1: 211 (1994)).

Dactylorhiza incarnata var. cruenta ×
D. majalis subsp. majalis

(*Dactylorhiza* ×*aschersoniana* (Hausskn.) Borsos & Soó, *Ann. Univ. Sci. Budapest. Rolando Eotvos, Sect. Biol.* iii. (Geobot. Monogr. Orch. Pannon. & Karpat. Fl. IV: 113 (1960)).

Dactylorhiza maculata subsp. maculata ×
D. majalis subsp. majalis

(*Dactylorhiza* ×*townsendiana* (Rouy) Soó, *Nom. Nov. Gen.* Dactylorhiza: 10 (1962)).

Dactylorhiza maculata subsp. fuchsii ×
D. majalis subsp. lapponica

Dactylorhiza majalis × D. sambucina

(*Dactylorhiza* ×*ruppertii* (M.Schulze) Borsos & Soó, *Ann. Univ. Sci. Budapest. Rolando Eotvos, Sect. Biol.* iii. (Geobot. Monogr. Orch. Pannon. & Karpat. Fl. IV: 122 (1960)).

LEFT TO RIGHT, FROM TOP

Dactylorhiza incarnata subsp. *incarnata* × D. majalis subsp. *lapponica*
Switzerland, Schwyz
Switzerland, Schwyz

Dactylorhiza incarnata var. *cruenta* × D. majalis subsp. *majalis*
Switzerland, Tessin
Switzerland, Tessin

Dactylorhiza maculata subsp. *maculata* × D. majalis subsp. *majalis*
Switzerland, Schwyz
Switzerland, Schwyz

Dactylorhiza maculata subsp. *fuchsii* × D. majalis subsp. *lapponica*
Switzerland, Obwalden
Switzerland, Obwalden
Switzerland, Obwalden
Switzerland, Schwyz

Dactylorhiza majalis × D. sambucina
France, Lozère
France, Lozère

HYBRIDS

Gymnadenia conopsea × G. odoratissima
(*Gymnadenia* ×*intermedia* Peterm., *Fl. Bienitz*: 30 (1841)).

Neotinea lactea × N. tridentata

Neotinea tridentata × N. ustulata
(*Neotinea* ×*dietrichiana* (Bogenh.) H.Kretzschmar, *Orch. Gen. Anacamptis, Orchis, Neotinea*: 464 (2007)).
NOTE. Occasional hybrids found where they grow together.

Ophrys apifera × O. holosericea subsp. holosericea
(*Ophrys* ×*albertiana* E.G.Camus, *Bull. Soc. Bot. France* 38: 40, 43 (1891)).

Ophrys apifera × O. scolopax
(*Ophrys* ×*minuticauda* Duffort, *Bull. Vulg. Sc. Nat. Gers*, 27: ii (1902)).

Ophrys apifera × O. sphegodes subsp. aveyronensis
(*Ophrys* ×*corvey-bironii* J.-M.Lewin, *Orchidophile (Asnières)* 140: 18 (2000)).

Ophrys argolica × O. reinholdii

LEFT TO RIGHT, FROM TOP

Gymnadenia conopsea × G. odoratissima
Switzerland, Baselland
Switzerland, Baselland

Neotinea lactea × N. tridentata
Crete
Crete

Neotinea tridentata × N. ustulata
France, Var

Ophrys apifera × O. holosericea subsp. *holosericea*
Germany, Black Forest
France, Alsace

Ophrys apifera × O. scolopax
France, Aveyron

Ophrys apifera × O. sphegodes subsp. *aveyronensis*
France, Aveyron

Ophrys argolica × O. reinholdii
Greece, Pelopponese
Greece, Pelopponese

Ophrys bertolonii × *O. holosericea*

Ophrys bertolonii × *O. insectifera*

Ophrys bertolonii × *O. scolopax*

Ophrys bertolonii × *O. tenthredinifera*
(*Ophrys* ×*sorrentinoi* Lojac., *Fl. Sicul.* 3: 41 (1909)).

Ophrys bombyliflora × *O. tenthredinifera*
(*Ophrys* ×*sommieri* G.Camus ex Cortesi, *Ann. Bot. (Rome)* 1: 360, in Iot. (1904)).

Ophrys ferrum-equinum × *O. reinholdii*
(*Ophrys* ×*kalteiseniana* B.Baumann & H.Baumann, *Mitteilungsbl. Arbeitskreis Heimische Orchid. Baden-Württemberg* 16(1): 127 (1984)).

Ophrys ferrum-equinum × *O. speculum*
(*Ophrys* ×*scalana* H.Baumann & Künkele, *Mitteilungsbl. Arbeitskreis Heimische Orchid. Baden-Württemberg* 18(3): 462 (1986)).

Ophrys ferrum-equinum × *O. sphegodes* subsp. *spruneri*
(*Ophrys* ×*mastii* P. Delforge, *Orchidophile* (Asnières) 59:497 (1983)).

LEFT TO RIGHT, FROM TOP

Ophrys bertolonii × *O. holosericea*
France (2)

Ophrys bertolonii × *O. insectifera*
France (2)

Ophrys bertolonii × *O. scolopax*
France (2)

Ophrys bertolonii × *O. tenthredinifera*
Italy, Puglia

Ophrys bombyliflora × *O. tenthredinifera*
Crete
Italy, Tuscany

Ophrys ferrum-equinum × *O. reinholdii*
Rhodes (8)
Rhodes (8)

Ophrys ferrum-equinum × *O. speculum*
Rhodes (8)

Ophrys ferrum-equinum × *O. sphegodes* subsp. *spruneri*
Greece (8)

Ophrys holosericea × O. insectifera

(*Ophrys* ×*devenesis* Rchb.f., *Icon. Fl. Germ. Helv.* 13-14: 87 (1851)).

Ophrys holosericea subsp. apulica × O. tenthredinifera

NOTE. *Ophrys holosericea* × *O. tenthredinifera* was described as *Ophrys* ×*maremmae* O. & E.Danesch, Ophrys *Hybr.*: 225 (1972)).

Ophrys holosericea subsp. grandiflora × O. umbilicata subsp. umbilicata

LEFT TO RIGHT, FROM TOP

Ophrys holosericea × *O. insectifera*
France (10)
France (8)
France (8)

Ophrys holosericea subsp. *apulica* × *O. tenthredinifera*
Italy, Puglia
Italy, Puglia

Ophrys holosericea subsp. *grandiflora* × *O. umbilicata* subsp. *umbilicata*
Cyprus (8)
Cyprus (8)

Ophrys holosericea subsp. *grandiflora* × *O. umbilicata* subsp. *lapethica*
Cyprus
Cyprus
Cyprus

Ophrys fusca × *O. lutea*
Crete
Tunisia (45)

HYBRIDS | 401

Ophrys holosericea* subsp. *grandiflora* × *O. umbilicata* subsp. *lapethica
(*Ophrys* ×*liebischiana* Kohlmüller, *Mitteilungsbl. Arbeitskreis Heimische Orchid. Baden-Württemberg* 22(4): 816 (1990)).

Ophrys fusca* × *O. lutea
(*Ophrys* ×*battandieri* E.G.Camus, *Monogr. Orchid.*: 307 (1908)).

Ophrys insectifera* subsp. *insectifera* × *O. sphegodes* subsp. *sphegodes
(*Ophrys* ×*hybrida* Pokorny ex Rchb.f. in Rchb., *Icon. Fl. Germ. Helv.* 13/14: 79 (1850-1851)).

Ophrys insectifera* subsp. *insectifera* × *O. sphegodes* subsp. *araneola

Ophrys insectifera* subsp. *aymoninii* × *O. sphegodes* subsp. *araneola

LEFT TO RIGHT, FROM TOP

Ophrys insectifera subsp. *insectifera* × *O. sphegodes* subsp. *sphegodes*
France, Aveyron
France, Aveyron
France, Aveyron
France, Aveyron
France, Aveyron

Ophrys insectifera subsp. *insectifera* × *O. sphegodes* subsp. *araneola*
France, Aveyron
France, Aveyron

Ophrys insectifera subsp. *aymoninii* × *O. sphegodes* subsp. *araneola*
France, Aveyron
France, Aveyron
France, Aveyron
France, Aveyron

Ophrys lunulata × Ophrys speculum
(*Ophrys ×syracusana* O.Danesch & E.Danesch, Ophrys *Hybr.*: 227 (1972)).

Ophrys lutea × O. scolopax
(*Ophrys ×pseudospeculum* DC., *Fl. France* ed. 3, 6: 332 (1815)).

Ophrys scolopax ×
O. sphegodes subsp. *araneola*

Ophrys scolopax subsp. *scolopax* ×
O. sphegodes subsp. *aveyronensis*
(*Ophrys ×bernardii* Looken, *Orchidophile (Asnières)* 75: 1211 (1987)).

LEFT TO RIGHT, FROM TOP

Ophrys lunulata × *O. speculum*
Sicily (8)
Sicily (8)

Ophrys lutea × *O. scolopax*
France, Tierques
France, Tierques
France, Tierques
France, Tierques

Ophrys scolopax subsp. *scolopax* ×
O. sphegodes subsp. *aveyronensis*
France, Aveyron
France, Aveyron
France, Aveyron

Ophrys scolopax subsp. scolopax × O. sphegodes subsp. sphegodes

(*Ophrys* ×*nouletii* E.G. Camus, *J. Bot.* (Morot) vii: 158 (1893)).

Ophrys scolopax subsp. heldreichii × O. tenthredinifera

(*Ophrys* ×*duvigneaudiana* P.Delforge & C.Delforge, *Orchidophile (Asnières)* 78: 1364 (1987)).

NOTE. *Ophrys scolopax × O. tenthredinifera* is *Ophrys* ×*composita* Pau, Le Monde des Plantes 30, 181: 1 (1929).

Ophrys sphegodes × O. tenthredinifera

(*Ophrys* ×*grampinii* Cortesi, *Ann. Bot. (Rome)* 1: 359 (1904)).

Orchis anthropophora × O. italica

(*Orchis* ×*bivonae* Tod., *Imparziale Giorn. Sc. Lett. Art.*: 34 (1840)).

NOTE. An occasional hybrid, varying in intensity of flower colour.

LEFT TO RIGHT, FROM TOP

Ophrys scolopax subsp. *scolopax* ×
O. sphegodes subsp. *sphegodes*
France, Aveyron
France, Aveyron
France, Aveyron
France, Aveyron
France, Aveyron
France, Aveyron

Ophrys scolopax subsp. *heldreichii* × *O. tenthredenifera*
Crete

Ophrys sphegodes × *O. tenthredinifera*
Italy, Tuscany
Italy, Tuscany

Orchis anthropophora × *O. italica*
Portugal, Coimbra
Portugal, Coimbra
Italy, Campania (45)

Orchis anthropophora × O. militaris
(*Orchis* ×*spuria* Rchb.f., *Bot. Zeitung (Berlin)* 7: 891 (1849)).

Orchis anthropophora × O. purpurea
(*Orchis* ×*macra* Lindl., *Syn. Brit. Fl., ed. 2*: 260 (1835)).
NOTE. A rare hybrid where the parents grow together.

Orchis anthropophora × O. simia
(*Orchis* ×*bergonii* Nanteuil, *Bull. Soc. Bot. France* 34: 422 (1888)).
NOTE. An occasional hybrid, varying in intensity of colour, some being very pale.

Orchis mascula × O. pallens
(*Orchis* ×*loreziana* Brügger, *Beitr. Kenntn. Ung. Chur.*: 58 (1874)).
NOTE. A rare hybrid where the parents coexist.

LEFT TO RIGHT, FROM TOP

Orchis anthropophora × *O. militaris*
France, Aveyron
Switzerland, Aargau
France, Massif Central

Orchis anthropophora × *O. purpurea*
France, Aveyron
France, Aveyron
France, Aveyron

Orchis anthropophora × *O. simia*
France, Massif Central
France, Massif Central
France, Massif Central

Orchis mascula × *O. pallens*
Germany, Black Forest
Germany, Black Forest
Germany, Black Forest

HYBRIDS

Orchis mascula × O. provincialis
(*Orchis* ×*penzigiana* A.Camus, *Iconogr. Orchid. Europe*: 270 (1928)).
NOTE. Occasionally found where the parents grow together.

Orchis militaris × O. purpurea
(*Orchis* ×*hybrida* (Lindl.) Boenn. ex Rchb., *Fl. Germ. Excurs.*: 125).
NOTE. Often found where the species grow together. Hybrid swarms resulting from backcrossing of primary hybrids to either parent are not infrequent.

Orchis militaris × O. simia
(*Orchis* ×*beyrichii* (Rchb.f.) A.Kern., *Verh. K.K. Zool.-Bot. Ges. Wien* 15: 208 (1865)).
NOTE. A not infrequent hybrid where the parents grow together.

Orchis purpurea × O. simia
(*Orchis* ×*angusticruris* Franch., *Fl. Loir.-et-Cher.*: 571 (1885)).
NOTE. An occasional hybrid.

Platanthera bifolia × P. chlorantha
(*Platanthera* ×*hybrida* Brügger, *Jahresber. Naturf. Ges. Graubündens* 2, 23-24: 118 (1880)).
NOTE. An occasional hybrid where the parents grow together.

LEFT TO RIGHT, FROM TOP

Orchis mascula × *O. provincialis*
France, Aveyron
France, Aveyron

Orchis militaris × *O. purpurea*
France, Aveyron
France, Aveyron
France, Gard
France, Hérault

Orchis militaris × *O. simia*
France, Occitanie
France, Occitanie

Orchis purpurea × *O. simia*
France, Languedoc-Roussillon
France, Aveyron

Platanthera bifolia × *P. chlorantha*
Switzerland, Baselland
Switzerland, Baselland

HYBRIDS

Serapias cordigera × S. lingua
(*Serapias* ×*ambigua* Rouy ex E.G.Camus, *Iconogr. Orch. Eur. Bass. Med.* 96 (1927–1929)).

LEFT TO RIGHT, TOP ROW
Serapias cordigera × S. lingua
France, Provence
France, Provence
Crete

INTERGENERIC HYBRIDS

Anacamptis laxiflora × Serapias vomeracea
(×*Serapicamptis rousii* (Du Puy) J.M.H.Shaw, *Orchid Rev. Suppl.*, 113, 1264: 20 (2005)).
NOTE. Occasionally found in S France.

Anacamptis pyramidalis × Gymnadenia conopsea
(×*Gymnanacamptis anacamptis* (F.Wilms.) Aesch. & Graebn., *Syn. Mitteleur. Fl.* iii: 855 (1907)).

Anacamptis pyramidalis × Gymnadenia odoratissima
(×*Gymnanacamptis odoratissima* Wildh., *Ber. Schweiz. Bot. Ges.* lxxvii: 433 (1967)).

Dactylorhiza incarnata × Gymnadenia conopsea
(×*Dactylodenia lebrunii* (E.G.Camus) Peitz, *Jahresber. Naturwiss. Vereins Wuppertal* 25: 190 (1972)).

LEFT TO RIGHT, FROM SECOND ROW
Anacamptis laxiflora × Serapias vomeracea
France, Var
France, Var

Anacamptis pyramidalis × Gymnadenia conopsea
Switzerland, Aargau
France, Aveyron
Switzerland, Aargau

Anacamptis pyramidalis × Gymnadenia odoratissima
France, Aveyron

Dactylorhiza incarnata × Gymnadenia conopsea
Switzerland, Schwyz
Switzerland, Schwyz

Dactylorhiza maculata subsp. maculata × Gymnadenia conopsea

(×*Dactylodenia legrandiana* (E.G.Camus) Peitz, *Jahresber. Naturwiss. Vereins Wuppertal* 25: 190 (1972)).

Dactylorhiza maculata subsp. fuchsii × Gymnadenia odoratissima

(×*Dactylodenia lawalreei* P.Delforge & D.Tyteca, *Bull. Roy. Bot. Belgique* 115, 2: 277 (1982)).

Gymnadenia conopsea × Nigritella miniata

(×*Gymnigritella godferyana* G.Keller, *Repert. Sp. Nov. Regni Veg. Sonderb.* A: 2, 270 (1933)).

Gymnadenia conopsea × Nigritella rhellicani

(×*Gymnigritella suaveolens* E.G.Camus, *J. Bot. (Morot)* vi: 484 (1892)).

Gymnadenia odoratissima × Nigritella rhellicani

(×*Gymnigritella heufleri* E.G.Camus, *J. Bot. (Morot)* vi: 484 (1892)).

Platanthera bifolia × Gymnadenia conopsea

(×*Gymplatanthera chodatii* (Lendn. ex Beauverd) E.G.Camus, Bergon & A.Camus, *Monogr. Orchid.*: 337 (1908)).

LEFT TO RIGHT, FROM TOP

Dactylorhiza maculata subsp. maculata × Gymnadenia conopsea
Switzerland, Schwyz

Dactylorhiza maculata subsp. fuchsii × Gymnadenia odoratissima
Switzerland, Grisons
Switzerland, Grisons

Gymandenia conopsea × Nigritella miniata
Switzerland, Grisons
Switzerland, Grisons

Gymnadenia conopsea × Nigritella rhellicani
Switzerland, Tessin
Switzerland, Grisons

Gymnadenia odoratissima × Nigritella rhellicani
Switzerland, Grisons
Switzerland, Grisons

Platanthera bifolia × Gymnadenia conopsea
France, Massif Central
France, Massif Central
France, Massif Central

Cypripedium calceolus
Italy, Dolomites (25)

Coeloglossum viride
Italy, Lazio (21)

ACKNOWLEDGEMENTS

We have been helped by many colleagues and friends in the preparation of this field guide. Our special thanks to Professor Richard Bateman for his Foreword and for giving us access to as yet unpublished analyses of selected European orchid genera. We are also most grateful to the late Thomas Renz, Samuel Sprunger, Werner Lehmann, Lukas Meyer and Roland Bühlmann and the Trustees of the Swiss Orchid Foundation for encouraging us from the inception of this project and for considerable help in sourcing photographs.

The following, in some cases represented by a surviving relative or copyright holder, have kindly allowed us to use their photographs where we lacked material (the number following each name identifies the images each has contributed): H. Baumann (1), P. Gölz (2), R. Lorenz (3), D. Rückbrodt (5), R. Peter (8), Basler Botanische Gesellschaft (9), K. Senghas (10), E. Bühler (11), A. Michel (12), R. Selig (13), H. Kretzschmar (14), W. Eccarius (15), R. Hansen (16), R. Hutchings (20), P. Harcourt-Davies (21), D. Turner-Ettlinger (22), D. Chesterman (23), D. Adelmann (24), Royal Botanic Gardens, Kew (26), E. Fischer (27), W. Baker (28), R. Bateman (30), E. Winter (31), R. van Vugt (40), L. Averyanov (46), N. Faurholdt (47), J.F. Christians (50), J. Renz (51), R. Piesl (52), A. Sands (53), K. Schaub (55), C.A.J. Kreutz (56), E. Gügel (57), B. Tattershall (58), A. Molnar (58), M. Benedetto (59), I. la Croix (61), J. Ross (62), H. Sundermann (63) and R.-B. Hansen (64). The authors photographs are labelled as follows: R. Kühn (unnumbered), P. Cribb (25) and H. Pedersen (45).

Gina Fullerlove and Lydia White of Kew Publishing have encouraged and guided us in the lengthy process of preparing the material for publication. Christine Beard and Nicola Thompson have used their considerable talent and patience to design the layout of the book while Georgie Hills has overseen the schedule and digitisation of the distribution maps. We appreciate immensely their help from the point of their enthusiastic acceptance of our proposal of the book through to its completion.

Orchis mascula subsp. *mascula*
England, Surrey (25)

APPENDIX 1: Keys to genera and species

A. Key to genera

1 Lateral sepals connate. Lip slipper-shaped, ovoid or urn-shaped, inflated. Median anther sterile, forming a shield-like staminode; lateral anthers fertile................ ***Cypripedium*** (key B)

Lateral sepals free. Lip without the above combination of traits. Median anther fertile, (sub) sessile; lateral anthers missing or vestigial.. 2

2 Lip ending in 4 long, thread-like processes *or* side lobes of the lip with conspicuously wavy margins.. ***Himantoglossum*** (key Q)

Lip not ending in 4 thread-like processes; side lobes, if any, not with conspicuously wavy margins.. 3

3 Inflorescence several- to many-flowered with the flowers spirally arranged in one or three rows .... ***Spiranthes*** (key H)

Inflorescence one- to many-flowered; flowers not spirally arranged.................................... 4

4 Lip cleft for at least its apical ¹/₃, glabrous, devoid of a spur; side lobes absent or only represented by a pair of tiny, papillose teeth at the very base of the lip .... ***Neottia*** (key E)

Lip cleft or not in its apical part; if cleft, always distinctly spurred at base, pubescent on the ventral surface and/or with well-developed side lobes 5

5 All leaves reduced to sheaths or scales 6

One or more well-developed leaves present (occasionally withered but still recognisable at the time of flowering)... 8

6 Lateral sepals violet to purple, more than 15 mm long. Ovary erect, more than 10 mm long .... ***Limodorum*** (key F)

Lateral sepals cream to yellow or green, sometimes brown in their distal part, less than 15 mm long. Ovary suberect to pendent, less than 10 mm long.. 7

7 Flowers spreading to suberect, resupinate. Lip with two short, smooth ridges at base, spurless... ***Corallorhiza*** (only one species, *C. trifida*)

Flowers pendent, non-resupinate. Lip with 4–6 long, verrucose lists and a 6–8 mm long spur .... ***Epipogium*** (only one species, *E. aphyllum*)

8 The apex of each sepal and petal forming a slender, club-like process..................................... .... ***Traunsteinera*** (key M)

Sepals and petals not forming slender, club-like processes 9

9 Lip provided with a constriction that divides it into a basal part (hypochile) and an apical part (epichile)... 10

Lip not differentiated in hypochile and epichile.. 13

APPENDIX: KEYS TO GENERA AND SPECIES

10   Leaf 1, basal, petiolate. Inflorescence 1-flowered (very rarely 2-flowered). Hypochile forming an apically cleft spur .... **Calypso** (only one species, *C. bulbosa*)
Leaves 2 or more (very rarely 1), alternate on the stem, sessile. Inflorescence few- to many-flowered. Hypochile spurless or with a spur that is not apically cleft.................................. 11

11   Bracts subsimilar to the sepals, longitudinally striated by conspicuous dark nerves. Column long-acuminate.. **Serapias** (key R)
Bracts markedly different from the sepals, not longitudinally striated. Column rounded...... 12

12   Flowers short-pedicellate, pendent to spreading. Epichile with 2–3 irregular calli at base. Anther yellow .... **Epipactis** (key D)
Flowers sessile, suberect to spreading. Epichile with 3–9 papillose or wavy longitudinal ridges. Anther white or purple.. **Cephalanthera** (key C)

13   Stem, sepals and ovaries densely glandular-pubescent .... **Goodyera** (key G)
Stem glabrous.. 14

14   Petals longer than sepals 15
Petals as long as or shorter than sepals 16

15   Leaves basal, lanceolate to ovate. Inflorescence with the flowers turning to all sides. Flowers honey-scented.. **Herminium** (only one species, *H. monorchis*)
Leaves alternate on the stem. Inflorescence secund. Flowers not honey-scented ....
.... **Gennaria** (only one species, *G. diphylla*)

16   Lip spurless.. 17
Lip spurred (spur sometimes very short) 22

17   Lip pubescent on (at least parts of) the ventral surface .... **Ophrys** (key P)
Lip glabrous 18

18   Leaves less than 0.4 cm broad.................................. **Chamorchis** (only one species, *C. alpina*)
Leaves more than 0.4 cm broad.. 19

19   Sepals and petals connivent to form a hood over the column. Lip deeply trilobed with deeply bilobulate mid-lobe.. **Orchis** p.p. (*O. anthropophora*)
Sepals and petals spreading or recurved. Lip entire 20

20   Bracts less than 0.3 times as long as the ovaries (including pedicels). Petals spreading. Lip recurved by a knee-like bend, rounded........................... **Liparis** (only one species, *L. loeselii*)
Bracts more than 0.3 times as long as the ovaries (including pedicels). Petals recurved. Lip straight, acute to cuspidate 21

21   Longest leaf more than 3 cm long, devoid of marginal bulbils. Lip cordate, cuspidate, uniformly pale green... **Malaxis** (only one species, *M. monophyllos*)
Longest leaf less than 3 cm long, often producing marginal bulbils in its distal part. Lip ovate, acute, pale green with 4 darker longitudinal bands..
.... **Hammarbya** (only one species, *H. paludosa*)

22   Flowers non-resupinate. Lip enveloping the column at base................................. 23
Flowers resupinate. Lip not enveloping the column... 24

APPENDIX: KEYS TO GENERA AND SPECIES 421

23 Lip tightly enveloping the column at base; spur less than 2 mm long ....
.... ***Nigritella*** (key K)
Lip loosely enveloping the column at base; spur at least 2 mm long ....
.... ***Gymnigritella*** (only one species, *G. runei*)

24 Lip chestnut to dark purple-brown proximally and green to yellow-green distally; spur apically cleft.. ***Steveniella*** (only one species, *S. satyrioides*)
Lip with a different colour scheme; spur not apically cleft 25

25 Lip green to yellow-green, deeply three-lobed with subsimilar, narrowly linear lobes. Fertile part of the stigma differentiated into two freely extending branches......................................
.... ***Habenaria*** (only one species, *H. tridactylites*)
Lip with a different combination of traits. Fertile part of the stigma entire, flat to concave 26

26 Inflorescence secund. Lateral sepals markedly falcate, more than 3.5 times as long as broad.. ***Ponerorchis*** (only one species, *P. cucullata*)
Inflorescence with the flowers turning to all sides (or nearly so). Lateral sepals oblique to slightly falcate, less than 3.5 times as long as broad.. 27

27 Petals more than half as broad as the lip. Viscidia 2, naked 28
Petals less than half as broad as the lip. Viscidia 1–2, contained in 1 or 2 bursicles 30

28 Lateral sepals connivent with the dorsal sepal and the petals to form a hood over the column. Lip trilobed with subacute lobes; spur slightly constricted at base ....
.... ***Pseudorchis*** (only one species, *P. albida*)
Lateral sepals widely spreading to reflexed. Lip entire to trilobed with broadly rounded lobes; spur not constricted at base... 29

29 Lip entire, at least 1.5 times as long as broad, white to (yellowish) green ....
.... ***Platanthera*** (key I)
Lip obscurely to distinctly trilobed, less than 1.5 times as long as broad, purple to rose-coloured or pure white .... ***Gymnadenia*** (key J)

30 Bracts not appressed to the ovaries 31
Bracts appressed to the ovaries... 32

31 Lip more than 1.5 times as long as broad; spur not distinctly longer than wide, appressed to the flat part of the lip .... ***Coeloglossum*** (only one species, *C. viride*)
Lip less than 1.5 times as long as broad (when flattened); spur distinctly longer than wide, not appressed to the flat part of the lip .... ***Dactylorhiza*** (key L)

32 Lip deeply (to moderately) trilobed, glabrous, adorned with numerous dots and/or dashes all over its ventral surface... ***Neotinea*** p.p. (key O)
Lip markings (if any) usually restricted to the central part of the lip; if distributed all over the ventral surface, the lip is entire to shallowly trilobed or the markings are in reality groups of tiny, coloured hairs.. 33

33 Spur 1–2 mm long.. ***Neotinea*** p.p. (key O)
Spur at least 3 mm long 34

APPENDIX: KEYS TO GENERA AND SPECIES

34 Flowers somewhat human-like, especially due to their deeply trilobed lips with arm-like side lobes and a bilobulate mid-lobe with leg- or skirt-like lobules. Lip ornamented with groups of tiny, coloured hairs (except in *O. italica* and *O. galilaea*)....................... *Orchis* p.p. (key N)
Flowers not human-like. Lip glabrous.. 35

35 All or nearly all leaves assembled in a basal rosette (at most 1 or 2 fastened higher up on the stem). Lateral sepals spreading to reflexed (in *O. spitzelii* sometimes loosely connivent with the petals and the dorsal sepal); never with conspicuous green nerves.............. *Orchis* p.p. (key N)
Leaves alternate on the stem or somewhat concentrated in a basal rosette (but still least 3 fastened higher up on the stem). Lateral sepals spreading to reflexed or connivent with the petals and the dorsal sepal to form a hood over the column; sometimes with conspicuous green nerves .... *Anacamptis* (key S)

B. Key to the species of *Cypripedium*

1 Flowers white, spotted with purple on lip, sepals and petals. Petals short and blunt. Lip urn-shaped .... *C. guttatum*
Flowers not spotted. Petals tapering to an acute tip. Lip ovoid to slipper-shaped 2

2 Flowers with a yellow lip and maroon sepals and petals. Petals usually spirally twisted......
.... *C. calceolus*
Flowers uniformly purple, pink or (rarely) white. Petals not spirally twisted ..... *C. macranthos*

C. Key to the species of *Cephalanthera*

1. Leaves (sub)erect, boat-shaped, loosely embracing the stem for most of their length (except sometimes the uppermost leaf). Flowers with a 1–4 mm long spur at the base of the lip 2
Leaves more or less spreading, flat to channelled (rarely slightly boat-shaped), not embracing the stem for most of their length. Flowers without a spur at the base of the lip 4

2 Flowers (creamy-)white. Sepals more than 25 mm long. Lip having a pointed epichile with 7–9 longitudinal ridges .... *C. epipactoides*
Flowers pink, purple, faintly rose-coloured, cream or white. Sepals at most 25 mm long; lip having an obtuse epichile with 4–7 longitudinal ridges.. 3

3 Flowers mainly pink to purple (very rarely white). Sepals 20–25 mm long. Petals subacute. Lip with a 3–4 mm long spur .... *C. kurdica*
Flowers faintly rose-coloured to (creamy-)white. Sepals 14–20 mm long. Petals obtuse to rounded. Lip with a 1–2 mm long spur .... *C. cucullata*

4 Leaves grey-green. Ovaries and inflorescence axis densely glandular-pubescent. Sepals and petals rose-pink (very rarely white). Lip pointed... *C. rubra*
Leaves mid-green. Ovaries and inflorescence axis (sub)glabrous. Sepals and petals white to cream (very rarely faintly rose-coloured). Lip rounded... 5

5 Plant with 6–13 foliage leaves, the longest usually more than 5 times as long as broad. Bracts of the uppermost flowers less than half as long as the ovaries .... *C. longifolia*
Plant with 1–6 foliage leaves, the longest usually less than 5 times as long as broad. Bracts of the uppermost flowers more than half as long as the ovaries... 6

6 Flowers (sub)erect, scarcely opening. Sepals 12–20 mm long.......................... *C. damasonium*

Flowers spreading (to suberect), opening to at least half-way. Sepals 20–30 mm long ....
.... *C. caucasica*

D. Key to the species of *Epipactis*

1 Aerial shoots usually forming extensive colonies (underground rhizome creeping). Epichile elastically attached to the hypochile... 2
Aerial shoots solitary or in small clusters (underground rhizome not creeping). Epichile firmly attached to the hypochile 3

2 Hypochile with obliquely triangular sidelobes; epichile suborbicular, white with a yellow spot... *E. palustris*
Hypochile devoid of side lobes; epichile (ovate-)oblong, brownish with a white tip............
.... *E. veratrifolia*

3 Internode between inflorescence and uppermost leaf at least twice as long as any other internode. Epichile bearing prominent, strongly wrinkled-tuberculate calli......................... 4
Internode between inflorescence and uppermost leaf usually less than twice as long as any other internode. Epichile bearing smooth to furrowed or verrucose calli............................ 5

4. Longest leaf more than 1.5 times as long as the internode above. Viscidium effective and lasting... *E. atrorubens*
Longest leaf less than 1.5 times as long as the internode above. Viscidium evanescent ....
.... *E. microphylla*

5. Vegetative parts glaucous-green to yellow-green. Inflorescence axis tomentose. Ovary pubescent to tomentose .... *E. condensata*
Vegetative parts occasionally yellow-green, but never glaucous-green. Inflorescence axis subglabrous to pubescent. Ovary glabrous to sparsely pubescent 6

6. Upper part of stem subglabrous (hairs few and short)....................................... *E. phyllanthes*
Upper part of stem distinctly pubescent... 7

7. Epichile longer than broad, acuminate.. *E. leptochila*
Epichile at least as broad as long, acute to rounded or emarginate 8

8. Ovary distinctly curved; pedicel more than 5 mm long .... *E. greuteri*
Ovary often oblique, but never distinctly curved; pedicel less than 5 mm long 9

9. Leaves strongly recurved. Inner lateral walls of the hypochile well separated at their junction with the epichile. Viscidium absent... *E. muelleri*
Leaves straight to lightly recurved. Inner lateral walls of the hypochile almost touching each other at their junction with the epichile. Viscidium absent or present 10

10. Longest leaf attached above the middle of the stem (inflorescence excluded). Epichile porrect, acute.. *E. albensis*
Longest leaf attached below the middle of the stem (inflorescence excluded). Epichile deflexed or recurved, usually obtuse to emarginate (rarely acute)....................... 11

11. Flowers opening to about half-way. Viscidium absent or ineffective.................................... 12
Flowers opening fully. Viscidium usually present and effective, very rarely evanescent or even absent... 13

12. Epichile smaller than hypochile, never with pink or rose tones. Viscidium present, but ineffective .... *E. pontica*

Epichile approximately as large as hypochile, usually with pink or rose tones. Viscidium absent.. *E. dunensis*

13. Leaves more or less distichous; longest leaf less than 1.5 times as long as the average internode length. Sepals and petals shining like silk .... *E. purpurata*

Leaves turning to all sides; longest leaf more than 1.5 times as long as the average internode length. Sepals and petals dull... *E. helleborine*

E. Key to the species of *Neottia*

1 The whole plant buff-coloured, lacking well-developed leaf laminas but carrying 2–5 leaf sheaths along the stem .... *N. nidus-avis*

The whole plant green (stem and flowers often suffused with reddish brown), carrying 2 sub-opposite leaves with well-developed laminas.. 2

2 Plant usually less than 20 cm tall. Leaves rhomboid to (triangular-)ovate, less than 3 cm long. Inflorescence axis subglabrous. Lip less than 5 mm long, with acuminate apical lobes and a pair of tiny teeth at base .... *N. cordata*

Plant usually more than 20 cm tall. Leaves elliptic, more than 3 cm long. Inflorescence axis densely glandular-pubescent. Lip more than 5 mm long, with blunt apical lobes and an entire base .... *N. ovata*

F. Key to the species of *Limodorum*

1 Lip divided into a long-spurred hypochile and an elliptic to cordate epichile with incurved sides.. *L. abortivum*

Lip not differentiated in hypochile and epichile, narrowly lanceolate, flat, spurless...............
.... *L. trabutianum*

G. Key to the species of *Goodyera*

1 Leaves less than 6 cm long. Inflorescence secund. Flower less than 6 mm long (excluding the pedicelled ovary) .... *G. repens*

Leaves more than 6 cm long. Inflorescence with the flowers turning to all sides (or nearly so). Flower more than 6 mm long (excluding the pedicelled ovary) ....
.... *G. macrophylla*

H. Key to the species of *Spiranthes*

1 Inflorescence with the flowers spirally arranged in three rows; bracts 10–17 mm long. Sepals at least 8 mm long.. *S. romanzoffiana*

Inflorescence with the flowers spirally arranged in one row; bracts 4–9 mm long. Sepals at most 7 mm long... 2

2 Inflorescence appearing after the basal rosette of spreading, ovate to elliptic leaves has withered; a new, young rosette usually sprouting at the base of the inflorescence. Lip white with a pale green central part.. *S. spiralis*

Inflorescence developing from the current year's basal rosette of (sub)erect, linear-lanceolate leaves. Lip white .... *S. aestivalis*

APPENDIX: KEYS TO GENERA AND SPECIES | 425

I. Key to the species of *Platanthera*

1 Spur distinctly longer than the ovary.. 2
  Spur shorter than to about as long as the ovary.. 7

2 Anther locules parallel or very nearly so... 3
  Anther locules distinctly diverging from each other towards the base................. 4

3 Anther locules very nearly touching each other .... *P. bifolia*
  Anther locules placed more than 1.3 mm apart... *P. muelleri*

4 Tip of spur neither bilaterally flattened nor club-shaped in side view................. *P. holmboei*
  Tip of spur bilaterally flattened, in side view slightly club-shaped........................ 6

6 Lateral sepals white to cream. Lip straight to lightly recurved .... *P. chlorantha*
  Lateral sepals green to yellow-green, sometimes rather pale. Lip moderately to strongly
  recurved .... *P. algeriensis*

7 Lateral sepals more than 7 mm long. Lip more than 7 mm long. Viscidia more than 2 mm
  apart.. *P. azorica*
  Lateral sepals less than 7 mm long. Lip less than 7 mm long. Viscidia less than 2 mm apart 8

8 Lip porrect, strongly incurved.. *P. micrantha*
  Lip porrect or pendent, straight to slightly recurved.. 9

9 Plant with 1–3 leaves (including small non-sheathing leaves). Inflorescence constituting
  less than 1/3 of the total plant height. Lip with 2 low, sometimes obscure longitudinal
  ridges .... *P. oligantha*
  Plant with more than 3 leaves (including small non-sheathing leaves). Inflorescence
  constituting more than 1/3 of the total plant height. Lip without ridges 10

10 Sheathing leaves linear-lanceolate to oblong-lanceolate. Lateral sepals lightly twisted,
  slightly shorter than the lip .... *P. hyperborea*
  Sheathing leaves ovate to obovate. Lateral sepals not twisted, slightly longer than
  the lip... *P. pollostantha*

J. Key to the species of *Gymnadenia*

1 Lip subentire to obscurely trilobed; spur less than 7 mm long................................. 2
  Lip obscurely to distinctly trilobed; spur more than 7 mm long 3

2 Inflorescence less than twice as long as wide. Flowers white with contrasting reddish column.
  Spur 1.5–3 mm long.. *G. frivaldii*
  Inflorescence more than twice as long as wide. Flowers pink to rose-coloured (rarely white)
  with whitish to yellowish column. Spur 3–6 mm long .... *G. odoratissima*

3 Plant usually less than 20 cm tall. Flowers with a clove-like scent. Lip obscurely (to distinctly)
  3-lobed .... *G. borealis*
  Plant usually more than 20 cm tall. Flowers with a sweet or carnation-like scent. Lip distinctly
  (to obscurely) 3-lobed 4

4 Broadest leaf usually less than 1.5 cm across. Inflorescence lax (to dense). Flowers with a
  faint, sweet to mouldy scent. Lip without distinct "shoulders"............................. *G. conopsea*
  Broadest leaf usually more than 1.5 cm across. Inflorescence dense. Flowers with a strong
  carnation-like scent. Lip often with distinct "shoulders" .... *G. densiflora*

426 | APPENDIX: KEYS TO GENERA AND SPECIES

K. Key to the species of *Nigritella*

1 Stem stout (upper part about twice as broad as the base of the lowermost bracts)... *N. buschmanniae*
Stem slender (upper part slightly broader than the base of the lowermost bracts)............. 2

2 Flower buds pink to rose-coloured... 3
Flower buds red to blackish red, purple-red or yellow... 5

3 Flowers weakly vanilla-scented. Individual sepals rose-coloured in their basal part, gradually turning white towards the tip.. *N. stiriaca*
Flowers without a vanilla-like scent. Individual sepals uniformly pink to pale rose-coloured.. 4

4 Inflorescence apically rounded throughout the flowering period. The constricted part of the lip (approximately $1/3$ above its base) c. 3 mm wide in its natural conformation...... *N. widderi*
Inflorescence apically pointed during the early phase of flowering (later rounded). The constricted part of the lip (approximately $1/3$ above its base) c. 2 mm wide in its natural conformation.. *N. lithopolitanica*

5 Flowers losing colour with age... *N. corneliana*
Flowers not losing colour with age.. 6

6 Flowers scentless to faintly aromatic... *N. gabasiana*
Flowers with a distinct scent of vanilla or chocolate 7

7 At least the uppermost flowers deep ruby-red. Lip with a saddle-like constriction approximately $1/3$ above its base... *N. miniata*
Uppermost flowers usually dark brownish red to blackish red (occasionally differently coloured, but never ruby-red). Lip without a saddle-like constriction 8

8 Bracts in the mid-portion of the inflorescence minutely papillose-dentate. Flowers vanilla-scented. Lip up to 6.5 mm long... *N. rhellicani*
Bracts in the mid-portion of the inflorescence (sub)entire. Flowers vanilla- or chocolate-scented. Lip more than 6.5 mm long.. *N. nigra*

L. Key to the species of *Dactylorhiza*

1 Lateral sepals of fully opened flowers connivent with the petals and the dorsal sepal to form a hood over the column.. *D. iberica*
Lateral sepals of fully opened flowers more or less spreading to reflexed........................... 2

2 Leaves unspotted. Lip trilobed with emarginate to truncate (rarely rounded) mid-lobe; lip markings, if any, never incorporating coherent loops... 3
Leaves spotted or unspotted. Lip entire or trilobed with acuminate to rounded (rarely truncate) mid-lobe; lip markings, if any, sometimes incorporating coherent loops................................ 6

3 Lip devoid of markings; spur upcurved... *D. romana*
Lip provided with markings (sometimes only few and confined to its extreme base); spur straight to downcurved.. 4

4 Longest leaf describing an adaxial angle of more than 55° relative to the stem. Flowers with yellow or purple-red (rarely salmon) ground colour. Spur more than 11 mm long, more than 1.4 times as long as the flat part of the lip... *D. sambucina*

Longest leaf describing an adaxial angle of less than 55° to the stem. Flowers with yellow ground colour. Spur less than 11 mm long, less than 1.4 times as long as the flat part of the lip 5

5   Usually more than 3 non-sheathing leaves below the inflorescence. Lip markings restricted to the basal part of the lip .... *D. insularis*
Usually less than 3 non-sheathing leaves below the inflorescence. Lip markings not restricted to the basal part of the lip .... *D. cantabrica*

6   Stem squeezable in its upper part (occasionally non-squeezable in *D. majalis* subsp. *lapponica, nieschalkiorum* and *traunsteinerioides*) 7
Stem non-squeezable in its upper part 11

7   Upper half of the stem (inflorescence not included) with at least as many nodes of sheathing leaves as the lower half 8
Upper half of the stem (inflorescence not included) with fewer nodes of sheathing leaves than the lower half 9

8   Plant usually less than 30 cm tall. Leaves usually marked with numerous small purplish-brown spots and streaks on both sides (less often unmarked). Lateral margins of the lip erose .... *D. euxina*
Plant usually more than 30 cm tall. Leaves unmarked. Lateral margins of the lip (sub) entire .... *D. armeniaca*

9   Bracts quite entire (i.e., marginal cells neither enlarged nor protruding) ...... *D. incarnata* p.p.
Bracts minutely crenate to serrate (i.e., marginal cells enlarged and protruding) 10

10   Lip of the lowermost flower (when flattened) more than twice as wide as the stem immediately below the inflorescence. Lateral sepals spreading (to nearly reflexed), more or less curved .... *D. majalis* p.p.
Lip of the lowermost flower (when flattened) less than twice as wide as the stem immediately below the inflorescence. Lateral sepals reflexed (erect), straight or nearly so .... *D. incarnata* p.p. (subsp. *cilicica*)

11   Leaves shiny green, unspotted. Lateral sepals of fully opened flowers only slightly spreading. Lip with obscure markings and widely spreading side lobes; spur distinctly thinner than the ovary .... *D. foliosa*
Plant without the above combination of traits 12

12   Lip moderately to deeply 3-lobed; mid-lobe approximately as large as side lobes ....
.... *D. maculata* p.p.
Lip slightly 3-lobed; mid-lobe distinctly smaller than side lobes 13

13   Spur not thicker than the ovary .... *D. maculata* p.p.
Spur thicker than the ovary 14

14   Leaves brown-spotted on the upper surface. Spur usually longer than the ovary ....
.... *D. urvilleana*
Leaves unspotted. Spur usually shorter than the ovary 15

15   Bracts longer than the flowers. Spur more than twice as thick as the ovary ....
.... *D. majalis* p.p. (subsp. *pythagorae*)

APPENDIX: KEYS TO GENERA AND SPECIES

Bracts shorter than the flowers. Spur less than twice as thick as the ovary ....
.... ***D. maculata*** p.p. (subsp. *maurusia*)

M. Key to the species of *Traunsteinera*

1 Flowers pink to rose-coloured with purple spots on the lip (and often also on the petals); very rarely white without markings. Lateral sepals usually less than 9 mm long.............. ***T. globosa***

Flowers white to cream without markings. Lateral sepals usually more than 9 mm long ....
.... ***T. sphaerica***

N. Key to the species of *Orchis*

1 Lip deeply trilobed with bilobulate mid-lobe; sometimes ornamented with groups of tiny, coloured hairs 2

Lip entire to deeply trilobed with entire to notched mid-lobe; never ornamented with groups of tiny, coloured hairs.. 9

2 Lip spurless; mid-lobe without a tooth in the sinus between the lobules..... ***O. anthropophora***

Lip with a well-developed spur; mid-lobe with a tiny to conspicuous tooth in the sinus between the lobules.. 3

3 Inflorescence flowering from the top downwards 4

Inflorescence flowering from the bottom upwards... 5

4 Lip ornamented with groups of tiny, pink to purple hairs; side lobes more than 5 times as long as broad, incurved in fully opened flowers.. ***O. simia***

Lip ornamented with (somewhat papillose) blackish-red spots; side lobes less than 5 times as long as broad, not incurved in fully opened flowers... ***O. galilaea***

5 Leaves with wavy margins. Lip glabrous; side lobes and the lobules of the mid-lobe pointed .... ***O. italica***

Leaves with flat margins. Lip ornamented with groups of tiny, coloured hairs; side lobes and the lobules of the mid-lobe not pointed 6

6 Lip with white to purple ground colour 7

Lip with yellow to cream or orange-brown ground colour... 8

7 Dorsal side of lateral sepals light (violet-)grey, never marbled. Mid-lobe of the lip long-clawed, its lobules usually less than twice as broad as the side lobes of the lip .... ***O. militaris***

Dorsal side of lateral sepals green to dark purple-brown or deep magenta, often marbled. Mid-lobe of the lip sessile to short-clawed, its lobules usually more than twice as broad as the side lobes of the lip.. ***O. purpurea***

8 Dorsal side of sepals yellow to yellow-green, sometimes suffused with brown. Lip with lemon-yellow to orange-brown ground colour; side lobes less than 3 times as long as broad, strongly falcate; lobules of the mid-lobe distinctly incurved in fully opened flowers...........
.... ***O. punctulata***

Dorsal side of sepals green to pale green, sometimes suffused with brown. Lip with cream to sulphur-yellow ground colour, sometimes with a rose hue; side lobes more than 3 times as long as broad, straight to weakly falcate; lobules of the mid-lobe not distinctly incurved in fully opened flowers .... ***O. adenocheila***

9 Lip with a cylindric to conical spur that is less than 4 times as long as wide....................... 10

APPENDIX: KEYS TO GENERA AND SPECIES | 429

Lip with a (sub)cylindric to clavate spur that is more than 4 times as long as wide 11

10 Spur descending .... *O. spitzelii*
Spur horizontal to ascending... *O. patens*

11 Dorsal sepal widely spreading; lateral sepals widely spreading, horizontally oriented. Spur thread-like, thinner than the ovary.. 12
Dorsal sepal more less incurved over the petals; lateral sepals spreading to reflexed, usually (sub)erect. Spur not thread-like, as thick as or thicker than the ovary.................................. 13

12 Leaves with or without brown spots. Side lobes of the lip obliquely oblong to obovate or fan-shaped; spur longer than the ovary .... *O. quadripunctata*
Leaves without brown spots. Side lobes of the lip linear; spur as long as or shorter than the ovary.. *O. brancifortii*

13 Spur distinctly tapering from base to apex.. 14
Spur not distinctly tapering from base to apex.. 15

14 Leaves with green ground colour. Lip shallowly to moderately 3-lobed with spreading to incompletely reflexed sides.. *O. anatolica*
Leaves with silvery grey-green ground colour. Lip deeply 3-lobed with reflexed sides.........
.... *O. sitiaca*

15 Central part of lip yellow, devoid of markings.. *O. pallens*
Central part of lip yellow, lilac, light purple or white, marked with a few to many red or purple dots (these occasionally weak or even absent, but never so in yellow-lipped species) 16

16 Sepals pale cream, making a sharp contrast to the golden-yellow lip .... *O. pauciflora*
Flowers with a different colour scheme (sometimes yellow, but without the indicated contrast) 17

17 Spur more than twice as long as the flat part of the lip.. *O. laeta*
Spur less than twice as long as the flat part of the lip.. 18

18 Lip longitudinally vaulted when viewed from the side 19
Lip not longitudinally vaulted when viewed from the side 20

19 Flowers predominantly sulphur-yellow (very rarely white or dirty purple) with distinct red dots on the central part of the lip.. *O. provincialis*
Flowers predominantly purple-red with obscure purple dots (if any) on the central part of the lip .... *O. mascula* p.p. (subsp. *laxifloriformis*)

20 Lip with spreading to slightly recurved side lobes; lip markings usually not extending from the central part of the lip to the mid-lobe (if they do, the spur is straight and often shorter than the ovary) .... *O. mascula* p.p.
Lip with recurved to reflexed side lobes; lip markings extending from the central part of the lip to the mid-lobe; spur upcurved, longer than the ovary .... *O. olbiensis*

O. Key to the species of *Neotinea*

1 Flower buds blackish red, making a conspicuous contrast to the predominantly white lips of already opened flowers .... *N. ustulata*
Flower buds white, green, rose-coloured or pink, not making a conspicuous contrast to the

430 | APPENDIX: KEYS TO GENERA AND SPECIES

predominantly white to pink lips of already opened flowers.. 2

2   Leaves with or without purple-brown spots. Inflorescence more or less secund. Sepals less than 5 mm long. Lip unspotted but sometimes marked with purple at base and with purple streaks on disc; spur less than half as long as the erect ovary............................... *N. maculata*

Leaves unspotted. Inflorescence with the flowers turning to all sides. Sepals more than 5 mm long. Lip finely spotted all over; spur more than half as long as the spreading to suberect ovary 3

3   Inflorescence hemispherical to ovoid, rarely cylindrical. Sepals with pink to rose, rarely white ground colour (occasionally pale green at their very base). Lip with more or less incurved side lobes.. *N. tridentata*

Inflorescence cylindrical. Sepals with pale green to greenish white ground colour. Lip with more or less recurved side lobes.. *N. lactea*

P.   Key to the species of *Ophrys*

Note: In the below key, reference is made to one or two partly stabilised hybrid complexes for any species that is believed to be involved in such complexes. In such cases, the descriptions and photos of the tentatively identified species and the relevant hybrid complex (or complexes) should be compared for final identification.

1   Column rounded.. 2

Column acute (to obtuse) or caudate 9

2   Stigmatic cavity approximately as wide as the anther.. 3

Stigmatic cavity approximately twice as wide as the anther.. 4

3   Dorsal sepal distinctly boat-shaped, strongly incurved. Petals elliptic to ovate-triangular. Lip with a conspicuous fringe of long, red-brown hairs; speculum almost completely covering the mid-lobe .... *O. speculum*

Dorsal sepal nearly flat, straight. Petals linear. Lip short-pubescent except on the speculum which covers less than half of the mid-lobe .... *O. insectifera*

4   Dorsal sepal more or less reflexed. Petals hairy. Lip with a terminal appendage or short point (sometimes hidden underneath the lip) 5

Dorsal sepal not reflexed. Petals glabrous. Lip devoid of a terminal appendage or short point.. 6

5   Sepals green to yellowish green. Lip deeply trilobed in its basal part, without a prominent tuft of hairs close to the tip.. *O. bombyliflora*

Sepals rose-purple to white with a green mid-nerve. Lip entire, with a prominent tuft of hairs close to the tip... *O. tenthredinifera*

6   Petals recurved, with wavy margins. Lip saddle-shaped with a slender, stalk-like base; mid-lobe hardly longer than the side lobes .... *O. atlantica*

Petals spreading to porrect, with (nearly) even margins. Lip straight to downcurved, sessile; mid-lobe distinctly longer than the side lobes 7

7   Lip without a basal furrow along its mid-line; speculum ending in a distinct (in subsp. *hayekii* often obscure), white to blue, ω-shaped band............. *O. omegaifera* (see also *O. ×brigittae*)

Lip with a distinct basal furrow along its mid-line; speculum not ending in a distinct, ω-shaped band (though often more brightly coloured in front).. 8

8   Lip margins spreading or upcurved; side lobes in most subspecies mainly yellow .... *O. lutea*

Lip with downcurved margins; side lobes mainly or exclusively brown ....
.... *O. fusca* (see also *O. ×brigittae*)

9 Column with a long, S-curved tip; pollinia with flaccid stalks.................................. *O. apifera*
   Column with a short, acute (to obtuse) tip; pollinia with firm stalks 10

10 Lip terminally with a broad and conspicuous, rectangular to rhomboid or obtriangular, often dentate appendage... 11
   Lip terminally with (rarely without) a small, triangular to awl-shaped (very rarely subrectangular) process 13

11 Lip entire (to shallowly trilobed); protuberances weakly developed to obliquely conical, distinctly isolated from the lip margin (i.e., side lobes flat, if recognisable).............................
   .... *O. holosericea* (see also *O. ×arachnitiformis*, *O. ×vicina*)
   Lip deeply trilobed; side lobes converted into obliquely conical to horn-shaped (or rarely less prominent) protuberances... 12

12 Dorsal sepal flat to shallowly boat-shaped, more or less incurved, from the base describing an obtuse angle to the column (so, consequently, not forming a roof over the latter). Petals spreading to slightly recurved, pubescent with spreading hairs...................................
   .... *O. scolopax* (see also *O. ×delphinensis*, *O. ×vicina*)
   Dorsal sepal more or less boat-shaped, strongly incurved, from the base parallel to the column (so, consequently, forming a roof over the latter). Petals recurved, pubescent with appressed hairs .... *O. umbilicata*

13 Lip markedly saddle-shaped; terminal process erect. Column (in side view) tapering towards the base; stigmatic cavity distinctly longer than wide ....
   .... *O. bertolonii* (see also *O. ×flavicans*)
   Lip straight; terminal process porrect or pointing downwards. Column (in side view) not tapering towards the base; stigmatic cavity approximately as long as wide 14

14 Petals linear-lanceolate. Lip deeply trilobed; side lobes and lateral parts of mid-lobe reflexed (making the lip appear entire and narrowly oblong when seen from above); mid-lobe with a broad yellow to light brown margin; speculum never framed with a white border, in most cases isolated from the lip base.. *O. lunulata*
   Without the above combination of traits.. 15

15 Basal part of lip stalk-like, with a pair of low, longitudinal ridges. Stigmatic cavity 1–1.5 times as wide as the anther .... *O. cilicica*
   Basal part of lip not stalk-like, devoid of ridges. Stigmatic cavity approximately twice as wide as the anther.. 16

16 Speculum isolated from the base of the lip or only connected to the base by delicate lines.. 17
   Speculum connected to the base of the lip by distinct, broad bands *or* speculum absent....
   19

17 Lip deeply trilobed; side lobes converted into (often weakly developed) protuberances; speculum sometimes white.. *O. reinholdii*
   Lip entire to moderately trilobed; protuberances, if any, distinctly isolated from the lip margin; speculum never white 18

18 Petals subglabrous. Lip with purplish-black to dark purplish-brown ground colour; basal part dark velvety.. *O. ferrum-equinum*
Petals hairy. Lip with (reddish-)brown to yellow-brown or olive-green ground colour; basal part with long, white to light brown hairs, especially in the marginal zones.......................
.... *O. argolica* (see also *O.* ×*delphinensis*)

19 Lip entire to moderately (rarely deeply) trilobed near the middle; protuberances, if any, distinctly isolated from the lip margin (i.e., side lobes flat, if recognisable)............................
.... *O. sphegodes* (see also *O.* ×*arachnitiformis*, *O.* ×*flavicans*)
Lip moderately to deeply trilobed near the base; side lobes converted into hump-shaped to obliquely conical protuberances... 20

20 Sepals uniformly rose to rose-purple (rarely white). Petals less than 3 mm long. Lip less than 8 mm long, distinctly shorter than the dorsal sepal.. *O. schulzei*
Sepals green to (purplish-)brown, often bicolored or variegated. Petals more than 3 mm long. Lip more than 8 mm long, as long as or longer than the dorsal sepal........................ 21

21 Petals at least 3 times as long as broad. Stigmatic cavity with white ground colour ....
.... *O. cretica*
Petals less than 3 times as long as broad. Stigmatic cavity with dark grey ground colour ...
.... *O. kotschyi*

Q. Key to the species of *Himantoglossum*

1 Lip obtriangular to fan-shaped; side lobes thread-like; mid-lobe divided into thread-like tails .... *H. comperianum*
Lip not obtriangular to fan-shaped; lobes not thread-like... 2

2 Lip mid-lobe not twisted, less than twice as long as the side lobes... 3
Lip mid-lobe more or less twisted, several times as long as the side lobes 4

3 Leaves assembled in a basal rosette. Lip highly variable in colour, usually with an olive-green marginal zone... *H. robertianum*
Leaves distributed along the stem. Lip pink with a pale centre, devoid of any green coloration.. *H. metlesicsianum*

4 Lip mid-lobe less than 30 mm long, less than 5 times as long as the side lobes, spatulate-linear, slightly notched at the tip .... *H. formosum*
Lip mid-lobe more than 30 mm long, more than 5 times as long as the side lobes, narrowly linear, entire to deeply cleft at the tip... 5

5 Lip base devoid of markings... *H. caprinum*
Lip base with purple markings 6

6 Inflorescence dense (flowers in the mid-portion usually placed less than 5 mm apart). Lip mid-lobe entire or slightly notched at the tip (incision shorter than 5 mm).................... *H. hircinum*
Inflorescence lax (flowers in the mid-portion usually placed more than 5 mm apart). Lip mid-lobe cleft at the tip (incision longer than 5 mm) 7

7 Lateral sepals less than 12 mm long. Lip without distinct "shoulders" at the base...............
.... *H. adriaticum*
Lateral sepals more than 12 mm long. Lip with distinct "shoulders" at the base 8

APPENDIX: KEYS TO GENERA AND SPECIES | 433

8   Floral bracts slightly longer than the ovaries. Lip base with circular to longitudinally elongated (rarely transverse) markings; side-lobes not reflexed; mid-lobe (brownish-) purple... *H. calcaratum*
    Floral bracts around twice as long as the ovaries. Lip base with transversely elongated (rarely circular) markings; side lobes reflexed; mid-lobe light green to olive-green.......... .... *H. montis-tauri*

R.   Key to the species of *Serapias*

1   Base of hypochile with one entire to emarginate and longitudinally furrowed callus......... 2
    Base of hypochile with two parallel or divergent keels.. 4

2   Callus of the hypochile (sub)entire; epichile pinkish purple to pale ochre-yellow, subglabrous in its basal part .... *S. lingua*
    Callus of the hypochile retuse to emarginate in front and furrowed along the mid-line; epichile reddish-brown to dark purplish-brown, distinctly hairy in its basal part 3

3   Epichile (ovate-)lanceolate, more than 6.5 mm broad..................................... *S. olbia*
    Epichile linear-triangular to linear-lanceolate, less than 6.5 mm broad.............. *S. strictiflora*

4   Labellum reddish-brown with a conspicuous white margin...................................... *S. nurrica*
    Labellum uniformly coloured 5

5   Flowers very small (epichile less than 11 mm long)................................. *S. parviflora*
    Flowers not very small (epichile more than 11 mm long)... 6

6   Flowers white to whitish-green; autogamous (pollinia disintegrating soon after anthesis, all ovaries swelling before the end of flowering)... *S. perez-chiscanoi*
    Flowers reddish-brown to dark purplish-brown, salmon-coloured or ochre-yellow (very rarely white to whitish-green); allogamous (pollinia not disintegrating, usually no or few ovaries swelling before the end of flowering)... 7

7   Epichile (ovate-)lanceolate, its maximum width less than 1.7 times the width of the junction between epichile and hypochile.. 8
    Epichile ovate to ovate-elliptic, its maximum width more than 1.7 times the width of the junction between epichile and hypochile... 9

8   Inflorescence lax at peak flowering (any bract in the mid-portion of the spike overlapping with the bract immediately above for less than half of its length, if at all) .... *S. vomeracea*
    Inflorescence dense at peak flowering (any bract in the mid-portion of the spike overlapping with the bract immediately above for more than half of its length) .... *S. orientalis*

9   Stem and leaf bases usually with purplish-brown dots and lines. Hypochile tunnel- to U-shaped in front view, with two divergent keels at base.................................... *S. cordigera*
    Stem and leaves without purplish-brown markings. Hypochile funnel-shaped in front view, with two parallel keels at base.. *S. neglecta*

S.   Key to the species of *Anacamptis*

1   Flowers opening (and withering) from the top of the inflorescence downwards................. 2
    Flowers opening (and withering) from the base of the inflorescence upwards................... 3

2   Flowers usually with purplish-pink ground colour. Lip shallowly trilobed; side lobes broader than and nearly as long as the mid-lobe .... *A. **boryi***

Flowers usually with white to rose ground colour. Lip deeply trilobed; side lobes narrower and distinctly shorter than the mid-lobe .... *A. **israelitica***

3   Lip bearing a pair of distinct calli at the base; spur thread-like........................ *A. **pyramidalis***

Lip devoid of distinct calli; spur sac-like or (conical-)cylindric 4

4   Lip (sub)entire.. 5

Lip trilobed, sometimes shallowly so... 7

5   Lip with recurved to reflexed sides, never with well-defined markings (but sometimes multicoloured with gradual transitions between colours); spur sac-like, its mid-portion distinctly thicker than the ovary.. *A. **collina***

Lip with spreading to incurved sides, with or without distinct markings; spur (conical-) cylindric, its mid-portion not thicker than the ovary 6

6   Lip with a conspicuous purple spot on each side of the spur mouth and usually with a row of minor markings along the mid-line.. *A. **cyrenaica***

Lip unspotted or with spots/lines arranged in a way that does not match the above scheme .... *A. **papilionacea***

7   All leaves distributed along the stem. Lateral sepals spreading to reflexed........................... 8

Most leaves assembled in a basal rosette. Lateral sepals connivent with the dorsal sepal and the petals to form a hood over the column 9

8   Lip mid-lobe shorter than or as long as the side lobes, rounded to truncate, sometimes retuse.. *A. **laxiflora***

Lip mid-lobe slightly longer than the side lobes, bilobulate (to emarginate)....... *A. **palustris***

9   Sepals with distinct green or purple veins; dorsal sepal rounded. Spur ascending above the ovary, often broadest at the tip .... *A. **morio***

Sepals without distinct green or purple veins; dorsal sepal acute to caudate. Spur descending, more or less parallel to the ovary, tapering to the tip... 10

10   Dorsal sepal acute to somewhat acuminate. Lip abruptly downcurved at the base, usually with purple spots.. *A. **coriophora***

Dorsal sepal long-acuminate to caudate. Lip not abruptly downcurved at the base, devoid of markings.. *A. **sancta***

Anacamptis pyramidalis
Greece, Pelion (25)

Ophrys tenthredinifera
Greece, Peloponnese (25)

APPENDIX 2: Bibliography

LITERATURE CITED

Bateman, R. M. 2020. Implications of next-generation sequencing for the systematics and evolution of the terrestrial orchid genus *Epipactis*, with particular reference to the British Isles. *Kew Bulletin* 75: 4 (22 pp.).

Brandrud, M. K., Baar, J., Lorenzo, M. T., Athanasiadis, A., Bateman, R. M., Chase, M. W., Hedrén, M. & Paun, O. 2020. Phylogenomic relationships of diploids and the origins of allotetrapoids in *Dactylorhiza* (Orchidaceae). *Systematic Biology* 69: 91–109.

Brandrud, M. K., Paun, O., Lorenz, R., Baar, J. & Hedrén, M. 2019. Restriction-site associated DNA sequencing supports a sister group relationship of *Nigritella* and *Gymnadenia* (Orchidaceae). *Molecular Phylogenetics and Evolution* 136: 21–28.

Delforge, P. 2016. *Orchidées d'Europe, d'Afrique du Nord et du Proche-Orient*. 4th ed. Delachaux et Niéstle, Paris.

Devey, D. S., Bateman, R. M., Fay, M. F. & Hawkins, J. A. 2008. Friends or relatives? Phylogenetics and species delimitation in the controversial European orchid genus *Ophrys*. *Annals of Botany* 101: 385–402.

Givnish, T. J., Spalink, D., Ames, M., Lyon, S. P., Hunter, S. J., Zuluaga, A., Iles, W. J. D., Clements, M. A., Arroyo, M. T. K., Leebens-Mack, J., Endara, L., Kriebel, R., Neubig, K. M., Whitten, W. M., Williams, N. H. & Cameron, K. M. 2015. Orchid phylogenomics and multiple drivers of their extraordinary diversification. *Proceedings of the Royal Society B* 282: 20151553 (10 pp.).

Govaerts, R., Bernet, P., Kratochvil, K., Gerlach, G., Carr, G., Alrich, P., Pridgeon, A.M., Pfahl, J., Campacci, M. A., Holland Batista, D., Tigges, H., Shaw, J., Cribb, P., George, A., Kreutz, K. & Wood, J. (2023). Plants of the world online: Orchidaceae. Facilitated by the Royal Botanic Gardens, Kew. Published on the internet; http://powo.science.kew.org/.

Greuter, W 2008. On the correct name of the Late Spider Orchid, and its appropriate spelling: *Ophrys holosericea*. *Journal Europäischer Orchideen* 40: 657–662.

Hedrén, M. & Tyteca, D. 2020. On the hybrid origin of *Dactylorhiza brennensis* and implications for the taxonomy of allotetraploid *Dactylorhiza*. *Journal Europäischer Orchideen* 52: 33–64.

Kühn, R., Pedersen, H. Æ. & Cribb, P. 2019. *Field guide to the orchids of Europe and the Mediterranean*. Kew Publishing, Royal Botanic Gardens, Kew.

BIBLIOGRAPHY

Pridgeon, A. M., Cribb, P. J., Chase, M. W. & Rasmussen, F. N. 2001. *Genera orchidacearum* 2. Oxford University Press, Oxford.

Ramírez, S. R., Gravendeel, B., Singer, R. B., Marshall, C. R. & Pierce, N. E. 2007. Dating the origin of the Orchidaceae from a fossil orchid with its pollinator. *Nature* 1042–1045.

Sramkó, G., Paun, O., Brandrud, M. K., Laczkó, L., Molnár V. A. & Bateman, R. M. 2019. Iterative allogamy-autogamy transitions drive actual and incipient speciation during the ongoing evolutionary radiation within the orchid genus *Epipactis* (Orchidaceae). *Annals of Botany* 124: 481–497.

ADDITIONAL SELECTED SYSTEMATIC PUBLICATIONS

Bateman, R. M & Denholm, I. 2013. Taxonomic reassessment of the British and Irish tetraploid marsh orchids. *New Journal of Botany* 2: 37–55.

Bateman, R. M., Hollingsworth, P. M., Preston, J., Luo, Y.-B., Pridgeon, A. M. & Chase, M. W. 2003. Molecular phylogenetics and evolution of Orchidinae and selected Habenariinae (Orchidaceae). *Botanical Journal of the Linnean Society* 142: 1–40.

Bateman, R. M., Molnár V. A. & Sramkó, G. 2017. *In situ* morphometric survey elucidates the evolutionary systematics of the Eurasian *Himantoglossum* clade (Orchidaceae: Orchidinae). *PeerJ* 5: e2893.

Bateman, R. M. & Rudall, P. J. 2013. Systematic revision of *Platanthera* in the Azorean archipelago: not one but three species, including arguably Europe's rarest orchid. *PeerJ* 1: e218.

Bateman, R. M., Sramko, G. & Paun, O. 2018. Integrating restriction site-associated DNA sequencing (RAD-seq) with morphological cladistic analysis clarifies evolutionary relationships among major species groups of bee orchids. *Annals of Botany* 121: 85–105.

Bellusci, F., Pellegrino, G., Palermo, A. M. & Musacchio, A. 2008. Phylogenetic relationships in the orchid genus *Serapias* L. based on noncoding regions of the chloroplast genome. *Molecular Phylogenetics and Evolution* 47: 986–991.

Breitkopf, H., Onstein, R. E., Cafasso, D., Schlüter, P. M. & Cozzolini, S. 2015. Multiple shifts to different pollinators fuelled rapid diversification in sexually deceptive *Ophrys* orchids. *New Phytologist* 207: 377–389.

Campbell, V. V., Rowe, G., Beebee, T. J. C. & Hutchings, M. J. 2007. Genetic differentiation amongst fragrant orchids (*Gymnadenia conopsea* s.l.) in the British Isles. *Botanical Journal of the Linnean Society* 155: 349–360.

Cribb, P. 1997. *The genus* Cypripedium. Timber Press, Portland, Oregon & Royal Botanic Gardens, Kew.

Eccarius, W. 2016. *Die Orchideengattung* Dactylorhiza. W. Eccarius, Eisenach.

Eccarius, W. 2022. *Die Orchideengattung* Gymnadenia *mit einem Exkurs zur Gattung* Pseudorchis. W. Eccarius, Eisenach.

Hedrén, M. 2001. Systematics of the *Dactylorhiza euxina/incarnata/maculata* polyploid complex in Turkey: evidence from allozyme data. *Plant Systematics and Evolution* 229: 23–44.

Hedrén, M., Lorenz, R., Teppner, H., Dolinar, B., Giotta, C., Griebl, N., Hansson, S., Heidtke, U., Klein, E., Perazza, G., Ståhlberg, D. & Surina, B. 2017. Evolution and systematics of polyploid *Nigritella* (Orchidaceae). *Nordic Journal of Botany* 2018: e01539.

Hedrén, M., Nordström, S., Persson Hovmalm, H. A., Pedersen, H. Æ. & Hansson, S. 2007. Patterns of polyploid evolution in Greek marsh orchids (*Dactylorhiza*: Orchidaceae) as revealed by allozymes, AFLPs, and plastid DNA data. *American Journal of Botany* 94: 1205–1218.

Kretzschmar, H., Eccarius, W. & Dietrich, H. 2007. *The orchid genera* Anacamptis, Orchis, Neotinea. Echinomedia Verlag, Bürgel.

Pedersen, H. Æ. 2006. Systematics and evolution of the *Dactylorhiza romana/sambucina* polyploid complex (Orchidaceae). *Botanical Journal of the Linnean Society* 152: 405–434.

Pedersen, H. Æ. & Faurholdt, N. 2007. Ophrys: *the bee orchids of Europe*. Royal Botanic Gardens, Kew.

Pridgeon, A. M., Cribb, P. J., Chase, M. W. & Rasmussen, F. N. 1999–2009. *Genera orchidacearum* 1–5. Oxford University Press, Oxford.

Reinhammar, L.-G. 1998. Systematics of *Pseudorchis albida s.l.* (Orchidaceae) in Europe and North America. *Botanical Journal of the Linnean Society* 126: 363–382.

Sardaro, L. M. S., Atallah, M., Picarella, M. E., Aracri, B. & Pagnotta, M. A. 2012. Genetic diversity, population structure and phylogenetic inference among Italian orchids of the *Serapias* genus assessed by AFLP molecular markers. *Plant Systematics and Evolution* 298: 1701–1710.

Sramkó, G., Molnar, V. A., Hawkins, J. A. & Bateman, R. M. 2014. Molecular phylogeny and evolutionary history of the Eurasiatic orchid genus *Himantoglossum s.l.* (Orchidaceae). *Annals of Botany* 114: 1609–1626.

RECOMMENDED REGIONAL ORCHID FLORAS AND EXCURSION GUIDES

Antonopoulos, Z. & Tsiftsis, S. 2017. *Atlas of the Greek orchids II*. Mediterraneo Editions, Rethymno.

Baumann, H., Blatt, H., Dierssen, K., Dietrich, H., Dostmann, H., Eccarius, W., Kretzschmar, H., Kühn, H.-D., Möller, O., Paulus, H. F., Stern, W. & Wirth, W. 2005. *Die Orchideen Deutschlands*. Verlag der Arbeitskreise Heimische Orchideen Deutschlands, Uhlstädt-Kirchhasel.

Bournérias, M. & Prat, D. (eds.) 2005. *Les orchidées de France, Belgique et Luxembourg*. 2nd ed. Biotope, Mèze.

Cole, S. & Waller, M. 2020. *Britain's orchids. A field guide to the orchids of Great Britain and Ireland*. Princeton University Press, Princeton.

Dusak, F. & Prat, D. (eds.) 2010. *Atlas des Orchidées de France*. Biotope, Mèze & Muséum National d'Histoire Naturelle, Paris.

Eccarius, W. & Dietrich, H. 2013. *Orchideen Wanderungen in Thüringen*. EchinoMedia Verlag Dr Kerstin Ramm, Albersdorf.

Griebl, N. 2015. *Orchideenwanderungen in Österreich*. Leopold Stocker Verlag, Graz & Stuttgart.

Griebl, N. 2017. *Orchideenparadiese Europas. Die schönsten Orchideenziele von Schweden bis Zypern*. Freya Verlag, Linz.

Harrap, A. & Harrap, S. 2005. *Orchids of Britain and Ireland. A field and site guide*. A & C Black, London.

Kretzschmar, H., Kretzschmar, G. & Eccarius, W. 2001. *Orchideen auf Rhodos. Ein Feldführer durch die Orchideenflora der "Insel des Lichts"*. H. Kretzschmar, Bad Hersfeld.

Kretzschmar, H., Kretzschmar, G. & Eccarius, W. 2002. *Orchideen auf Kreta, Kasos und Karpathos. Ein Feldführer durch die Orchideenflora der zentralen Inseln der Südägäis*. H. Kretzschmar, Bad Hersfeld.

Kreutz, C. A. J. 1998. *Die Orchideen der Türkei*. C. A. J. Kreutz, Landgraaf.

Kreutz, C. A. J., Fateryga, A. V. & Ivanov, S. P. 2018. *Orchids of the Crimea*. Kreutz Publishers, Sint Geertruid.

Kreutz, K., Shifman, A., Schot, R. & Talmon, Y. 2021. *Orchids of Israel*. Kreutz Publishers, Sint Geertruid.

Martin, R., Vela, E. & Ouni, R. 2015. Orchidées de Tunisie. *Bulletin de la Société Botanique du Centre-Ouest*, new ser., special no. 44: 1–160.

Parker, S. 2014. *Wild orchids of the Algarve: how, when and where to find them*. First Nature, Bwlchgwyn.

Reinhard, H. R., Gölz, P., Peter, R. & Widermuth, H. 1991. *Die Orchideen der Schweiz und angrenzender Gebiete*. Fotorotar, Egg.

Tsiftsis, S. & Antonopoulos, Z. 2017. *Atlas of the Greek orchids I*. Mediterraneo Editions, Rethymno.

Vakhrameeva, M. G., Tatarenko, I. V., Varlygina, T. I., Torosyan, G. K. & Zagulskii, M. N. 2008. *Orchids of Russia and adjacent countries (within the borders of the former USSR)*. A. R. G. Gantner Verlag, Ruggell.

Wartmann, B. A. 2020. *Die Orchideen der Schweiz. Der Feldführer*. 3rd ed. Haupt Verlag, Bern.

Wartmann, B. A. & Wartmann, C. 2018. *Orchideenwanderungen: 24 Routen zu Hotspots in der Schweiz*. Haupt, Bern.

Neotinea tridentata and *Ophrys fusca* subsp. *fusca*
Greece, Crete

INDEX OF SCIENTIFIC NAMES

Accepted names in roman, synonyms in *italic*. **Bold** numbers indicate main entry.
Underlined numbers indicate a photograph or image.

A

Aceras 27, 177
  hircinum 342
  maculatum 207
Amesia,
  microphylla 55
  palustris 51
ANACAMPTIS 2, 10, 25, 27, 33, **362**, 422
  ×alata 390
  boryi **380**, 381, 434
  brancifortii 196
  champagneuxii 385
  collina 13, **368**, 369, 434
  coriophora 6, **372**, 434
    subsp. coriophora **372**, 373
    subsp. fragrans 371, **372**, 373, 374
  cyrenaica **375**, 434
  iberico 371
  israelitica **380**, 381, 434
  Key to the species of Anacamptis **433**
  ×laniccae 390
  laxiflora 33, **362**, 390, 410, 434
    subsp. dielsiana 364, **365**
    subsp. dinsmorei 364, **365**
    subsp. laxiflora **362**, 363
    × A. morio 390, 391
    × Serapias vomeracea 390, **410**, 411
  longicornu 387
  morio 14, 16, 22, **383**, 390, 434
    subsp. *albanica* 384
    subsp. caucasica **384**, 388
    subsp. champagneuxii **385**, 387
    subsp. longicornu vi, 385, 386, **387**
    subsp. morio 382, **383**
    subsp. picta **388**, 389
    subsp. syriaca **388**, 389
    × A. papilionacea 390, 391
    × A. pyramidalis 390, 391
  ×nicodemi 390
  palustris **366**, 434
    subsp. elegans 366, 367
    subsp. palustris **366**, 367
    subsp. robusta **368**, 369

  papilionacea 368, **370**, 371, **376**, 390, 434
    subsp. expansa 378, **379**
    subsp. palaestina 378, **379**
    subsp. papilionacea x, **376**, 377
    subsp. *rubra* 376
    subsp. *schirwanica* 376
    subsp. *thaliae* 376
  picta 388
  pyramidalis 6, 16, 362, **370**, 371, 390, 410,
    434, 435
    var. *brachystachys* 370
    var. *cerigensis* 370
    var. *nivea* 370
    var. *serotina* 370
    var. *tanayensis* 371
    × Gymnadenia conopsea **410**, 411
    × Gymnadenia odoratissima **410**, 411
  sancta 371, 434
  syriaca 388
  urvilleana 371
Androrchis,
  pallens 196
  pauciflora 200
  provincialis 199
  quadripunctata 195
  sitiaca 191
  spitzelii subsp. *cazorlensis* 188
Anteriorchis 362
  coriophora 372
    var. *carpetana* 372
    var. *martrinii* 372
  fragrans 372
  sancta 371
Arachnites 214
  bertolonii 271
  biancae 288
  bombyliflorus 221
  fuciflorus 283
  fuscus 228
  lunulatus 271
  luteus 225
  tenthredinifer 223
Arthrochilium veratrifolium 48

444 | INDEX OF SCIENTIFIC NAMES

B

Barlia 27, 339
  canariensis 191
  metlesicsiana 341
  robertiana 339
  spitzelii 186

C

CALYPSO 2, 25, **85**, 420
  borealis 85
  bulbosa 1, <u>84</u>, **85**, 420
CEPHALANTHERA 2, 8, 14, 25, **42**, 48, 420
  caucasica 14, **46**, <u>47</u>, 422
  conferta 45
  cucullata <u>44</u>, **45**, 422
    subsp. epipactoides 42
    subsp. *floribunda* 45
    subsp. *kurdica* 45
  damasonium 19, **46**, <u>47</u>, 422
    subsp. *kotschyana* 46
  epipactoides **42**, <u>43</u>, 45, 422
  Key to the species of Cephalanthera **422**
  kotschyana 46
  kurdica 14, <u>44</u>, **45**, 422
  latifolia 46
  longifolia 19, <u>44</u>, **45**, 422
    ochroleuca 46
  pallens 45, 46
  rubra 19, **42**, <u>43</u>, 422
CHAMORCHIS 25, **174**, 420
  alpina 14, <u>16</u>, **174**, <u>175</u>, 420
COELOGLOSSUM ix, 25, 28, **126**, 421
  diphyllum 95
  viride 1, 19, 28, <u>**126**</u>, <u>127</u>, 421
Comperia 27, 339
  comperiana 339
CORALLORHIZA **86**, 419
  trifida 2, **86**, <u>87</u>, 419
CYPRIPEDIUM 2, 3, <u>4</u>, 8, 25, <u>38</u>, 419, 422
  bulbosum 85
  calceolus 1, <u>8</u>, <u>19</u>, 23, **38**, <u>39</u>, 41, 422
  guttatum 1, 38, <u>40</u>, **41**, 422
  Key to the species of Cypripedium **422**
  macranthos 1, 38, <u>40</u>, **41**, 422
  ×ventricosum 41

D

×Dactylodenia (Dactylorhiza × Gymnadenia) 33,
  390
  lawalreei 413
  lebrunii 410
  legrandiana 413
Dactylorchis 128

DACTYLORHIZA ix, 8, 9, 20, <u>21</u>, 25, 28, 31, 32, 33,
  126, **128**, **129**, 421
  alpestris 152
  amblyoloba 151
  andoeyana 148
  armeniaca 128, <u>144</u>, **145**, 427
  ×aschersoniana 393
  atlantica 169
  baltica 152
  battandieri 151
  baumanniana 165
    subsp. *smolikana* 165
  cantabrica 128, <u>136</u>, **137**, 427
  caramulensis 147
  ×carnea 390
  carpatica 160
  coccinea 145
  cordigera 155
    subsp. *bosniaca* 155
    subsp. *pindica* 165
  cruenta 140
  curvifolia 160
  devillersiorum 160
  ebudensis 172
  elata 169
    subsp. *ambigua* 169
    subsp. *atlantica* 169
    subsp. *sesquipedalis* 169
    var. *algerica* 169
    var. *durandii* 169
    var. *elongata* 169
    var. *iberica* 169
    var. *occitanica* 169
    var. *sesquipedalis* 169
  ericetorum 147
  euxina 128, **138**, <u>139</u>, 427
    subsp. *armeniaca* 145
  flavescens 134
  foliosa <u>146</u>, **147**, 170, 427
  fuchsii 148
    subsp. *okellyi* 148
    subsp. *psychrophila* 148
  graeca 155
  guimaraesii 134
  iberica <u>128</u>, **129**, 426
  incarnata 20, 128, **138**, <u>139</u>, 390, 393, 410, 427
    subsp. *baumgartneriana* 138
    subsp. cilicica **142**, <u>143</u>, 427
    subsp. coccinea <u>144</u>, **145**
    subsp. *cruenta* 140
      × D. majalis subsp. majalis <u>392</u>, **393**
    subsp. incarnata,
      var. cruenta **140**, <u>141</u>, 393
      var. incarnata **138**, <u>139</u>, **140**, <u>141</u>, 142

INDEX OF SCIENTIFIC NAMES 445

var. ochroleuca **142**, 143
subsp. *jugicrucis* 138
subsp. *lobelii* 145
subsp. *pulchella* 138
var. *baumgartneriana* 138
var. *drudei* 138
var. *dunensis* 145
var. *haussknechtii* 138
var. *hyphaematodes* 138
var. *jugicrucis* 138
var. *kotschyi* 142
var. *lobelii* 145
var. *macrophylla* 138
× D. maculata **390**, 391
× D. majalis subsp. lapponica 392, **393**
× Gymnadenia conopsea **410**, 411
insularis 128, 136, **137**, 427
islandica 147
kalopissii 158
subsp. *macedonica* 162
kerryensis 164
var. occidentalis 164
Key to the species of Dactylorhiza **426**
kolaensis 147
lapponica 160
subsp. *angustata* 160
×legrandiana 33, 390
libanotica 132
macedonica 162
maculata 128, 146, **147**, 390, 427, 428
subsp. *battandieri* 151
subsp. *caramulensis* 147
subsp. fuchsii 16, 33, **148**, 149, 393, 413
× D. majalis subsp. lapponica 392, **393**
× Gymnadenia odoratissima 412, **413**
subsp. hebridensis 148
subsp. *islandica* 147
subsp. maculata 9, 146, **147**, **148**, 149, 393, 413
× D. majalis subsp. majalis 392, **393**
× Gymnadenia conopsea 412, **413**
subsp. maurusia 150, **151**, 428
subsp. *podesta* 147
subsp. *pyrenaica* 147
subsp. saccifera 150, **151**
subsp. *savogiensis* 147
subsp. *sudetica* 147
subsp. *transsilvanica* 147
var. *elodes* 147
var. *podesta* 147
var. *rhoumenis* 147
var. *transsilvanica* 147
majalis 33, 128, **152**, **153**, 427
subsp. baltica **152**, 153, **154**, 155
subsp. brennensis 170

subsp. calcifugiens 154, **155**
subsp. cordigera 154, **155**
subsp. elata 170
subsp. elatior **156**, 157
subsp. integrata **156**, 157, 158, 159, 170
subsp. kalopissii **158**, 159, 162
subsp. lapponica 160, 161, 172, 393, 427
subsp. macedonica 155, **162**
subsp. majalis **152**, 153, 164, 393
subsp. nieschalkiorum **163**, 427
subsp. occidentalis 152, **164**
subsp. pindica **165**
subsp. purpurella **166**, 167
var. cambrensis **166**, 167
var. maculosa **166**, 167
var. pulchella **166**, 167
subsp. pythagorae 168, **169**, 427
subsp. sesquipedalis 168, **169**
subsp. sphagnicola 155, **170**, 171
subsp. traunsteinerioides 160, **172**, 173, 427
× D. sambucina 392, **393**
nieschalkiorum 163
ochroleuca 142
osiliensis 156
osmanica 142
subsp. *anatolica* 142
phoenissa 151
pindica 165
pontica 172
praetermissa 156
subsp. schoenophila 158
var. *integrata* 156
var. *junialis* 156, 157
psychrophila 148
pulchella 138
purpurella 166
pythagorae 169
romana 14, 128, **132**, 133, 426
subsp. georgica **134**, 135
subsp. guimaraesii **134**, 135
subsp. romana **132**, 133
×ruppertii 393
ruthei,
subsp. osiliensis 156
subsp. ruthei 155
saccifera 151
subsp. *bithynica* 151
subsp. *gervasiana* 151
sambucina 14, 128, **130**, **131**, 137, 426
savogiensis 147
schurii 160
smolikana 165
sphagnicola 170
sudetica 147

INDEX OF SCIENTIFIC NAMES

×townsendiana 393
traunsteineri 160
   subsp. *bohemica* 160
   subsp. *carpatica* 160
   subsp. *curvifolia* 160
   subsp. *irenica* 160
   subsp. *turfosa* 160
traunsteinerioides 172
   subsp. *francis-drucei* 172
   var. *francis-drucei* 172
umbrosa 142
urvilleana 128, **172**, <u>173</u>, 427
   subsp. *ilgazica* 172
viridis 28, 126

E

EPIACTIS ix, 2, <u>4</u>, 7, 8, 14, 19, 25, 31, **48**, 420
albensis **73**, <u>74</u>, 423
aspromontana 56
atropurpurea var. *tremolsii* 62
atrorubens <u>6</u>, <u>8</u>, 19, <u>50</u>, **51**, 55, 423
   subsp. atrorubens <u>50</u>, **51**, 52
   subsp. parviflora **52**, <u>53</u>
bithynica 58
bugacensis 70
calabrica 56
campeadorii 70
cardina 62
collaris 56
condensata <u>54</u>, **55**, 423
confusa 74
cretica 74, <u>75</u>
cucullata 45
degenii 56
densifolia 62
distans 56
dunensis 20, <u>70</u>, **71**, 424
   var. *tynensis* 70
etrusca 56
exilis 74
fageticola 74
fibri 73
greuteri **66**, <u>67</u>, 423
   var. *preinensis* 66
halacsyi 65
helleborine 1, 19, 31, **56**, <u>57</u>, 424
   subsp. bithynica **58**, <u>59</u>
   subsp. *condensata* 55
   subsp. helleborine 31, **56**, <u>57</u>, **58**, <u>59</u>
   subsp. *latina* 62
   subsp. *leptochila* 69
   subsp. *muelleri* 73
   subsp. neerlandica <u>60</u>, **61**
     var. neerlandica **61**

     var. renzii <u>60</u>, **61**
   subsp. *phyllanthes* 74
   subsp. *renzii* 61
   subsp. tremolsii **62**, <u>63</u>
   var. castanearum 62
   var. *microphylla* 55
   var. *neerlandica* 61
   var. *orbicularis* 56
   var. *youngiana* 61
heraclea 62
Key to the species of Epipactis **423**
kleinii 52
krymmontana 55
kuenkeleana 55
latifolia 56
   subsp. *purpurata* 65
   var. *microphylla* 55
   var. *muelleri* 73
leptochila **68**, **69**, 423
   subsp. leptochila <u>68</u>, **69**
   subsp. neglecta **68**, **69**
   var. *dinarica* 69
   var. *futakii* 69
   var. *komoricensis* 66
   var. *neglecta* 69
   var. *peitzii* 69
   var. *savelliana* 69
   var. *thesaurensis* 69
levantina 62
lucana 56
lusitanica 62
maestrazgona 69
maricae 70
mecsekensis 73
meridionalis 56
microphylla <u>54</u>, **55**, 423
molochina 56
moravica 73
muelleri **73**, <u>74</u>, 423
   subsp. *cerritae* 73
   var. *dunensis* 70
   var. *leptochila* 69
   var. *saltuaria* 73
nauosaensis 66
neerlandica 61
   var. *renzii* 61
neglecta 69
nordeniorum 73
olympica 66
palustris 9, 20, 31, <u>34</u>, 48, <u>50</u>, **51**, 423
parviflora 52
persica 74
   subsp. *gracilis* 74
   subsp. *pontica* 57

INDEX OF SCIENTIFIC NAMES | **447**

phyllanthes **74**, <u>75</u>, 423
  var. *degenera* 74
  var. *olarionensis* 74
  var. *pendula* 74
  var. *vectensis* 74
pinovica 69
placentina 73
  var. *robatschiana* 73
pontica **56**, <u>57</u>, 424
provincialis 70
pseudopurpurata 65
purpurata 19, <u>64</u>, **65**, 424
  subsp. halacsyi <u>64</u>, **65**
  subsp. purpurata <u>64</u>, **65**
  subsp. rechingeri **66**, <u>67</u>
  var. *pollinensis* 65
rechingeri 66
renzii 61
rhodanensis 70
sancta 70
sanguinea 56
schubertiorum 56
spiridonovii 51
subclausa 51
tallosii 73
taurica 74
tremolsii 62
troodi 74
turcica 62
veratrifolia 31, **48**, <u>49</u>, 423
viridiflora 65
voethii 56
EPIPOGIUM <u>4</u>, 25, **80**, 419
  aphyllum 1, 2, 3, 24, 35, **80**, <u>81</u>, 419
    f. *albiflorum* 80
  generalis 80

G

GENNARIA 25, 95, 420
  diphylla **95**, 420
GOODYERA 2, 25, **89**, 420
  Key to the species of Goodyera **424**
  macrophylla 1, <u>88</u>, **89**, 424
  repens 1, <u>88</u>, **89**, 424
GYMNADENIA ix, 25, 28, 31, 33, 96, **110**, 114, 124, 421
  albida subsp. *straminea* 99
  archiducis-joannis 121
  austriaca 118
    var. *gallica* 118
  bifolia 100
    var. *bifolia* 100
  borealis <u>112</u>, **113**, 425
  buschmanniae 122

carpatica 117
conopsea viii, 6, <u>16</u>, 28, 110, <u>112</u>, **113**, 114, 124, 410, 413, 425
  subsp. *borealis* 113
  subsp. *densiflora* 110
  var. *borealis* 113
  × Gymnadenia odoratissima **<u>394</u>**, <u>395</u>
  × Nigritella miniata <u>412</u>, **413**
  × Nigritella rhellicani <u>412</u>, **413**
corneliana 121
densiflora 6, 19, **110**, <u>111</u>, 113, 425
dolomitensis 122
frivaldii **110**, <u>111</u>, 425
gabasiana 114
×intermedia 394
Key to the species of Gymnadenia **425**
lithopolitanica 117
miniata 122
nigra subsp. *austriaca* 118
odoratissima 6, **114**, <u>115</u>, 394, 410, 413, 425
  × Nigritella rhellicani <u>412</u>, **413**
rhellicani 117
rubra 122
  var. *stiriaca* 124
runei 124
stiriaca 124
straminea 99
widderi 121
×Gymnanacamptis (Gymnadenia × Anacamptis) 33, 390
  anacamptis 410
  odoratissima 410
GYMNIGRITELLA 25, 28, **124**, 421
  runei 28, **124**, <u>125</u>, 421
×Gymnigritella,
  godferyana 413
  heufleri 413
  suaveolens 413
×Gymplatanthera chodatii 413

H

HABENARIA 25, **94**, 421
  albida var. *straminea* 99
  gymnadenia var. *borealis* 113
  longebracteata 106
  micrantha 109
  odoratissima 114
  tridactylites 1, **94**, 421
HAMMARBYA 2, 25, **82**, 420
  paludosa 3, 9, 20, <u>21</u>, **82**, 420
Helleborine,
  atrorubens 51
  leptochila 69
    var. *dunensis* 70

INDEX OF SCIENTIFIC NAMES

muelleri 73
palustris 51
persica 74
purpurata 65
rubra 42
viridiflora f. *dunensis* 70
HERMINIUM 25, 93, 420
monorchis 1, 6, 19, <u>92</u>, **93**, 420
Herorchis 362
boryi 380
champagneuxii 385
israelitica 380
longicornu 387
morio 382
picta 388
var. *albanica* 384
var. *caucasica* 384
var. *skorpili* 384
syriaca 388
HIMANTOGLOSSUM 14, 25, 27, 339, 419
adriaticum <u>**344**</u>, 345, 432
affine 345
bolleanum 345
calcaratum 345, **347**, 433
subsp. calcaratum <u>346</u>, **347**
subsp. *jankae* 347
subsp. rumelicum <u>346</u>, **347**, 348
caprinum <u>**345**</u>, 347, 348, 432
subsp. *bolleanum* 345
subsp. *levantinum* 345
subsp. *robustissimum* 347
subsp. *rumelicum* 347
comperianum 13, <u>338</u>, **339**, 432
formosum **342**, <u>343</u>, 432
galilaeum 345
hircinum 6, 16, 339, <u>**342**</u>, <u>343</u>, 344, 345, 347, 432
subsp. *calcaratum* 347
var. *caprinum* 345
jankae 347
var. *calcaratum* 347
var. *rumelicum* **347**
Key to the species of Himantoglossum **432**
metlesicsianum <u>**341**</u>, 432
montis-tauri 345, **348**, <u>349</u>, 433
robertianum 13, <u>338</u>, **339**, 340, 341, 432
samariense 345
satyrioides 336

I

Isias 348

J

Jonorchis abortiva 79

L

Leucorchis 96
albida 96
frivaldii 110
Limnorchis,
hyperborea 99
longebracteata 106
LIMODORUM 14, 25, 79, 419
abortivum 2, 13, 24, 35, <u>78</u>, **79**, 424
subsp. *trabutianum* 80
var. *trabutianum* 80
Key to the species of Limodorum **424**
rubrum 42
trabutianum 2, **80**, <u>81</u>, 424
LIPARIS 2, 25, **83**, 420
loeselii 9, 20, **83**, 420
Listera 27, 76, 78
cordata 76
ovata 76
Lonchitis 348
Loroglossum 339
formosum 342
Lysiella obtusata subsp. *oligantha* 100

M

MALAXIS 2, 25, **85**, 420
monophyllos 3, 9, 20, <u>84</u>, **85**, 420
paludosa 82
Monorchis herminium 93
Myodium 214

N

NEOTINEA 25, 27, **207**, 421
commutata 213
var. *angelica* 213
×dietrichiana 394
intacta 207
Key to the species of Neotinea **429**
lactea <u>210</u>, **211**, 213, 394, 430
var. *corsica* 211
× N. tridentata **394**, <u>395</u>
maculata <u>206</u>, **207**, 430
tridentata 211, **213**, 394, 430
subsp. conica <u>212</u>, **213**
subsp. tridentata <u>212</u>, **213**, 442
× N. ustulata <u>**394**</u>
ustulata 16, 19, **208**, 394, 429
f. *aestivalis* 208
var. aestivalis <u>**208**</u>, <u>209</u>
var. ustulata **208**, <u>209</u>
NEOTTIA 25, 27, **76**, 419
corallorhiza 86
cordata 20, **76**, <u>77</u>, 424

INDEX OF SCIENTIFIC NAMES 449

Key to the species of Neottia **424**
nephrophylla 76
nidus-avis <u>1</u>, 2, 6, 35, <u>78</u>, **79**, 424
ovata 6, 19, **76**, <u>77</u>, 424
Neottianthe 27, 96
cucullata 96
NIGRITELLA ix, 25, 28, **114**, 124, 421
archiducis-joannis 121
austriaca 118
bicolor 122
buschmanniae **122**, <u>123</u>, 426
corneliana <u>120</u>, **121**, 426
subsp. *bourneriasii* 121
gabasiana **114**, <u>115</u>, 426
globosa 174
hygrophila 122
Key to the species of Nigritella **426**
lithopolitanica <u>116</u>, **117**, 426
subsp. *corneliana* 121
miniata 14, **<u>122</u>**, <u>123</u>, 124, 413, 426
minor 122
nigra 28, **118**, <u>119</u>, 426
subsp. austriaca **118**, <u>119</u>
subsp. *corneliana* 121
subsp. *gallica* 118
subsp. *iberica* 118
subsp. nigra **118**, <u>119</u>, 124
rhellicani 14, <u>116</u>, **117**, 413, 426
rubra 122
runei 124
stiriaca **124**, <u>125</u>, 426
widderi <u>120</u>, **121**, 426

O

Odontorchis,
lactea 211
tridentata 213
ustulata 208
OPHRYS ix, 2, **5**, 6, 7, 10, 13, 14, 25, 30, 31, 32, **214**, 420
achillis 231
aegaea 275
subsp. *lucis* 280
aegirtica 283
aeoli 284
aesculapii 248
aestivalis 90
africana 228
alasiatica 265
×albertiana 394
algarvensis 236
alpina 174
amanensis 248
subsp. *antalyensis* 248

subsp. *iceliensis* 265
amphidami 223
×*anapei* 288
andria 284
annae 283
antalyensis 248
anthropophora 177
antiochiana 325
aphrodite 288
apifera **8**, 19, **240**, <u>241</u>, <u>242</u>, 243, 244, 394, 431
f. aurita **243**
f. austroalsatica <u>242</u>, **243**
f. bicolor <u>242</u>, **243**
f. botteronii 243
f. friburgensis <u>242</u>, **243**
f. trollii <u>242</u>, **243**
subsp. *botteronii* 243
subsp. *trollii* 243
var. *aurita* 243
var. *bicolor* 243
var. *botteronii* 243
var. *chlorantha* 240
var. *friburgensis* 243
var. *trollii* 243
× O. holosericea subsp. holosericea **394**, <u>395</u>
× O. scolopax **394**, <u>395</u>
× O. sphegodes subsp. aveyronensis **394**
appennina 283
appolonae 235
aprilia 223
apulica 287
arachnites var. *attica* 313
×arachnitiformis **332**, <u>333</u>, 431, 432
nm. "arachnitiformis" **332**, <u>333</u>
nm. "archipelagi" **<u>333</u>**
nm. "montis-leonis" **332**, <u>333</u>
nm. "splendida" **332**, <u>333</u>
aramaeorum 283
araneola 251
subsp. *cretensis* 257
aranifera 244
f. *epirotica* 257
subsp. *araneola* 251
subsp. *helenae* 258
subsp. *macedonica* 265
subsp. *renzii* 248
var. *atrata* 252
var. *taurica* 265
archimedea 227
archipelagi 333
argensonensis 251
argentaria 251
argolica **275**, 334, 335, 394, 432
subsp. aegaea <u>274</u>, **275**

INDEX OF SCIENTIFIC NAMES

subsp. argolica <u>274</u>, **275**, 334
subsp. *atargatis* 280
subsp. biscutella 276, <u>277</u>, 278
subsp. climacis **276**, <u>277</u>
subsp. crabronifera **278**
subsp. elegans **279**
subsp. lesbis **280**, <u>281</u>
subsp. lucis **280**, <u>281</u>, 325
subsp. *pollinensis* 278
× O. reinholdii **394**, <u>395</u>
ariadnae 318
arnoldii 228
aspea 227
astarte 313
astypalaeica 232
atlantica **240**, <u>241</u>, 430
subsp. *hayekii* 239
attaviria 228
attica 313
f. *flavomarginata* 317
subsp. *rhodia* 313
aurelia 329
ausonia 251
aveyronensis 252
aymoninii 217
balcanica 306
balearica 329
×battandieri 401
beloniae 318
benacensis 329
×bernardii 402
bertolonii 218, <u>270</u>, **271**, 327, 329, 397, 431
× O. holosericea <u>396</u>, **397**
× O. insectifera <u>396</u>, **397**
× O. scolopax <u>396</u>, **397**
× O. tenthredinifera <u>396</u>, **397**
bertoloniiformis 330
biancae 288
bicornis 306
biscutella 276
blitopertha 231
bombyliflora 13, **221**, 397, 430
f. *parviflora* 222
var. bombyliflora <u>220</u>, **221**
var. parviflora **222**
× O. tenthredinifera <u>396</u>, **397**
bornmuelleri 288
f. *grandiflora* 294
subsp. *grandiflora* 294
subsp. *ziyaretiana* 288
botteronii 243
bremifera 301
×brigittae **326**, 430, 431
nm. "sitiaca" **326**, <u>327</u>

nm. "vasconica" **326**, <u>327</u>
brutia 252
bucephala 314
caesiella 228
calliantha 291
calocaerina 228
caloptera 261
calypsus 335
candica 291
carduchorum 288
carmeli subsp. *attica* 313
castellana 332
catalaunica 329
caucasica 265
subsp. *cyclocheila* 265
celiensis 296
cephaloniensis 306
cerastes 306
cesmeensis 231
ceto 306
chalkae 336
chestermanii 291
chiosica 283
cilentana 244
ciliata 218
subsp. *lusitanica* 218
cilicica 14, 218, <u>324</u>, **325**, 431
cinereophila 232
cinnabarina 283
clara 231
classica 244
climacis 276
colossaea 283
×composita 405
conradiae 305
corallorhiza 86
corbariensis 301
cordata 76
cornuta 306
subsp. *heldreichii* 306
cornutula 306
corsica 225
×corvey-bironii 394
crabronifera 278
creberrima 228
cressa 228
cretensis 257
cretica **28**, **318**, <u>319</u>, <u>320</u>, <u>321</u>, 322, 325, 432
subsp. *ariadnae* 318, 321
subsp. *bicornuta* 318
subsp. cretica 321
subsp. *karpathensis* 318
creticola 228
creutzburgii 326

INDEX OF SCIENTIFIC NAMES 451

cyclocheila 265
cypria 322
cytherea 291
cythnia 227
dalmatica 271
delforgei 231
delmeziana 244
×*delphinensis* **334**, 431, 432
×*devenesis* 398
dianica 228
dicipulus 306
dictynnae 223
dinarica 287
dinsmorei 313
dodekanensis 336
druentica 283
drumana 329
×*duvigneaudiana* 405
dyris 236
elatior 292
elegans 279
eleonorae 232
ellinicaea 336
epirotica 257
episcopalis 283
    var. *samia* 296
eptapigiensis 228
exaltata 244
    subsp. *marzuola* 332
    subsp. *sundermannii* 276
explanata 330
fabrella 231
ferrum-equinum **272**, 397, 432
    subsp. *aegaea* 275
    subsp. *argolica* 275
    subsp. *climacis* 276
    subsp. ferrum-equinum **272**, <u>273</u>
    subsp. gottfriediana **272**, <u>273</u>
    subsp. *lesbis* 280
    subsp. *lucis* 280
    subsp. *spruneri* 264
    var. *pseudogottfriediana* 272
    × O. reinholdii O. <u>396</u>, **397**
    × O. speculum O. <u>396</u>, **397**
    × sphegodes subsp. spruneri O. <u>396</u>, **<u>397</u>**
ficuzzana 228
flammeola 228
×*flavicans* **327**, 431, 432
    nm. "balearica" <u>328</u>, **329**
    nm. "benacensis" <u>328</u>, **329**, 330
    nm. "bertoloniiformis" **330**
    nm. "catalaunica" <u>328</u>, **329**
    nm. "drumana" **<u>329</u>**
    nm. "explanata" **<u>330</u>**

    nm. "flavicans" **329**
    nm. "melitensis" **<u>330</u>**
    nm. "promontorii" **330**, <u>331</u>
    nm. "tarentina" **330**, <u>331</u>
flavomarginata 317
fleischmannii 236
friburgensis 243
fuciflora 283
    subsp. *andria* 284
    subsp. *apiformis* 305
    subsp. *apulica* 287
    subsp. *biancae* 288
    subsp. *bornmuelleri* 288
    subsp. *candica* 291
    subsp. *chestermanii* 291
    subsp. *elatior* 292
    subsp. *grandiflora* 294
    subsp. *lacaitae* 295
    subsp. *oblita* 296
    subsp. *oxyrrhynchos* 296
    subsp. *pallidiconi* 299
    subsp. *parvimaculata* 299
    subsp. *pollinensis* 278
    subsp. *tetraloniae* 300
    var. *ziyaretiana* 288
funerea 228
fusca **228**, 235, 326, 401, 431
    subsp. *atlantica* 240
    subsp. *blitopertha* <u>230</u>, **231**
    subsp. cinereophila **232**, <u>233</u>
    subsp. *durieui* 240
    subsp. *dyris* 236
    subsp. *fleischmannii* 236
    subsp. fusca **228**, <u>229</u>, <u>230</u>, 442
    subsp. *hayekii* 239
    subsp. iricolor **232**, <u>233</u>
    subsp. *omegaifera* 235
    subsp. pallida <u>234</u>, **235**
    var. *atlantica* 240
    var. *durieui* 240
    × O. lutea 399, **401**
gackiae 231
galilaea 227
    subsp. *melena* 228
garganica 261
    subsp. *passionis* 261
    subsp. *sipontensis* 263
gazella 228
gortynia 258
gottfriediana 272
    subsp. *elegans* 279
grammica 265
×*grampinii* 405
grigoriana 264

INDEX OF SCIENTIFIC NAMES

halia 283
hansreinhardii 265
hebes 244
heldreichii 236, 306
  var. *pseudoapulica* 335
  var. *scolopaxoides* 335
helenae 258
helios 283
hellenica 225
hespera 228
heterochila 336
hippocratis 306
hittitica 265
holosericea 16, 19, 30, **283**, 302, 332, 397, 431
  subsp. andria **284**
  subsp. apulica <u>286</u>, **287**, 296
    × O. tenthredinifera **398**, <u>399</u>
  subsp. biancae **288**, <u>289</u>
  subsp. bornmuelleri **288**, <u>289</u>
  subsp. candica 224, <u>290</u>, **291**
  subsp. chestermanii 224, <u>290</u>, **291**
  subsp. *cyrenaica* 283
  subsp. elatior **292**, <u>293</u>
  subsp. grandiflora **294**, 401
    × O. umbilicata subsp. lapethica 398, 399, **401**
    × O. umbilicata subsp. umbilicata **398**, <u>399</u>
  subsp. holosericea <u>29</u>, <u>282</u>, **283**, <u>285</u>, 296, 299, 394
  subsp. lacaitae **295**
  subsp. oblita **296**, <u>297</u>
  subsp. oxyrrhynchos 288, **296**, <u>297</u>
  subsp. pallidiconi <u>298</u>, **299**
  subsp. parvimaculata <u>298</u>, **299**
  subsp. *taloniensis* 283
  subsp. tetraloniae **300**
  subsp. *vanbruggeniana* 283
  var. gracilis 284
    × O. insectifera **398**, <u>399</u>
    × O. tenthredinifera 398
holubyana 336
homeri 335
hospitalis 232
×hybrida 390, 401
hygrophila 301
hystera 265
icariensis 272
iceliensis 265
illyrica 251
incantata 251
incubacea 252
  subsp. *pacensis* 252
  var. *dianensis* 252
  var. *septentrionalis* 252
insectifera <u>7</u>, 19, **214**, 397, 430

subsp. aymoninii <u>216</u>, **217**, 401
  × O. sphegodes subsp. araneola <u>400</u>, **401**
subsp. insectifera 7, 401
  × O. sphegodes subsp. araneola <u>400</u>, **401**
  × O. sphegodes subsp. sphegodes <u>400</u>, **401**
subsp. *philippi* 310
subsp. subinsectifera **217**
subvar. *aurita* 243
var. *apiformis* 305
var. *myodes* 214
iricolor 232
isaura 309
israelitica 239
janrenzii 265
×kalteiseniana 397
karadenizensis 301
kedra 231
Key to the species of Ophrys **430**
khuzestanica 301
knossia 265
konyana 325
korae 223
kotschyi **322**, <u>323</u>, 432
  subsp. *cretica* 318
kreutzii 312
kurdica 325
kurdistanica 325
labiosa 272
lacaena 283
lacaitae 295
lapethica 317
  subsp. *pamphylica* 317
latakiana 318
leochroma 223
lepida 227
leptomera 306
lesbis 280
leucadica 228
leucophthalma 265
levantina 294
  subsp. *grandiflora* 294
liburnica 244
ligustica 261
×liebischiana 401
lindia 228
litigiosa 251
loeselii 83
lojaconoi 228
lucana 228
lucentina 228
lucifera 228
lucis 280
luminosa 296
lunulata <u>270</u>, **271**, 402, 431

INDEX OF SCIENTIFIC NAMES | 453

× O. speculum **402**, <u>403</u>
lupercalis 228
luristanica 244
lusitanica 218
lutea **225**, 231, 401, 402, 431
   subsp. aspea <u>226</u>, **227**
   subsp. *funerea* 228
   subsp. galilaea <u>226</u>, **227**
   subsp. lutea **225**
   subsp. melena **228**, <u>229</u>
   subsp. *omegaifera* 235
   subsp. *quarteirae* 227
   var. *minor* 227
   × O. scolopax **402**, <u>403</u>
lycia 261
lyciensis 283
lycomedis 223
macedonica 265
magniflora 329
majellensis 244
malacitana 231
mammosa 265
   subsp. *gortynia* 258
   subsp. *mouterdeana* 265
   subsp. *posteria* 265
   subsp. *transhyrcana* 269
×*maremmae* 398
maritima 251
marmorata 228
marzensis 271
masticorum 306
×*mastii* 397
mattinatae 301
medea 283
melena 228
melitensis 330
meropes 231
mesaritica 228
minipassionis 261
minoa 291
minuscula 306
×*minuticauda* 394
mirabilis 239
monophyllos 85
monorchis 93
montis-leonis 333
morio 265
morisii 332
murgiana 263
muscifera 214
mycenensis 306
negadensis 244
nidus-avis 79
×*normanii* 224

×nouletii 405
numida 227
obaesa 228
oblita 296
oestrifera,
   subsp. *bremifera* 301
   subsp. *elbursana* 301
   subsp. *heldreichii* 306
   subsp. *latakiana* 318
   subsp. *orientalis* 313
   subsp. *philippi* 310
   subsp. *phrygia* 310
olympiotissa 272
omegaifera 231, <u>234</u>, **235**, 326, 430
   subsp. dyris **236**, <u>237</u>, 326
   subsp. fleischmannii **236**, <u>237</u>
   subsp. hayekii <u>238</u>, **239**
   subsp. israelitica <u>238</u>, **239**
   subsp. omegaifera <u>234</u>, **235**, 326
   var. *basilissa* 235
orphanidea 306
ortuabis 228
ovata 76
oxyrrhynchos 296
   subsp. *biancae* 288
   subsp. *lacaitae* 295
pallida 235
pallidula 231
paludosa 82
panattensis 332
panormitana 244
parosica 228
parvimaculata 299
parvula 228
passionis 261
pectus 235
pelinaea 326
penelopeae 225
peraiolae 228
perpusilla 231
persephonae 231
peucetiae 299
phaidra 231
pharia 287
phaseliana 228
×*philippi* 310
phryganae 225
phrygia 310
picta 305
pinguis 283
×*plakotiana* 258
×*poelmansiana* 244
pollinensis 278
polycratis 235

INDEX OF SCIENTIFIC NAMES

polyxo 306
posidonia 300
praemelena 228
promontorii 330
provincialis 247
proxima 228
pseudoatrata 261
pseudobertolonii 329
pseudomammosa 265
pseudomelena 227
×pseudospeculum 402
punctulata 228
regis-ferdinandii 221
reinhardiorum 322
reinholdii 13, **322**, 394, 397, 431
    subsp. *antiochiana* 325
    subsp. *leucotaenia* 325
    subsp. reinholdii **322**, <u>323</u>
    subsp. straussii <u>324</u>, **325**
renzii 248
rhodia 313
rhodostephane 306
riojana 251
riphaea 223
romolinii 271
rueckbrodtiana 231
sabulosa 231
saliarisii 283
samia 296
samiotissa 336
santonica 301
sappho 306
saratoi 329
×scalana 397
schulzei 14, **244**, <u>245</u>, 432
    var. *curdica* 244
scolopax <u>7</u>, 33, **301**, 314, 334, 335, 390, 394, 397, 402, 405, 431
    subsp. apiformis <u>304</u>, **305**
    subsp. conradiae <u>304</u>, **305**, 313
    subsp. cornuta **306**, <u>307</u>
    subsp. heldreichii **306**, <u>307</u>, <u>308</u>, 405
        × O. tenthredinifera <u>404</u>, **405**
    subsp. isaura <u>308</u>, **309**
    subsp. *jugurtha* 305
    subsp. philippi **310**, <u>311</u>
    subsp. phrygia **310**, <u>311</u>, 312
    subsp. rhodia <u>312</u>, **313**
    subsp. *sardoa* 305
    subsp. scolopax **301**, 302, <u>303</u>, 402, 405
        × O. sphegodes subsp. aveyronensis **402**, <u>403</u>
        × O. sphegodes subsp. sphegodes <u>404</u>, **405**
    var. *minutula* 301
    var. *philippi* 310

    var. *sepalina* 301
    × O. sphegodes subsp. araneola **402**
    × O. tenthredinifera **404**
scyria 326
sepioides 306
serotina 283
sicula 227
sintenisii 265
    subsp. *kotschyi* 322
    subsp. *straussii* 325
sitiaca 326
×sommieri 397
×sorrentinoi 397
spectabilis 223
speculum **218**, 397, 402, 430
    f. *regis-ferdinandii* 221
    subsp. lusitanica **218**, <u>219</u>
    subsp. regis-ferdinandii <u>220</u>, **221**
    subsp. speculum **218**, <u>219</u>
sphegifera 305
sphegodes 16, 19, 23, **244**, 327, 329, 332, 405, 432
    subsp. aesculapii **248**, <u>249</u>, 257
    subsp. amanensis **248**, <u>249</u>, 264
    subsp. araneola 26, <u>250</u>, **251**, 401, 402
    subsp. atrata **252**, <u>253</u>
    subsp. aveyronensis <u>7</u>, **252**, <u>253</u>, **254**, <u>255</u>, 394, 402
    subsp. *catalcana* 265
    subsp. cretensis <u>256</u>, **257**
    subsp. epirotica <u>256</u>, **257**
    subsp. *garganica* 261
    subsp. gortynia **258**, <u>259</u>
    subsp. *grassoana* 244
    subsp. helenae **258**, <u>259</u>
    subsp. *integra* 332
    subsp. *litigiosa* 251
    subsp. *lunulata* 271
    subsp. lycia <u>260</u>, **261**
    subsp. *mammosa* 258, 265
    subsp. passionis <u>260</u>, **261**, <u>262</u>
    subsp. *provincialis* 247
    subsp. *sintenisii* 265
    subsp. sipontensis **263**
    subsp. sphegodes **244**, <u>245</u>, <u>246</u>, 251, 401, 405
        var. provincialis **247**
    subsp. spruneri **264**, 272, 397
    subsp. taurica <u>265</u>, <u>266</u>, <u>267</u>, <u>268</u>
    subsp. *transhyrcana* 269
    var. *argentaria* 251
    var. transhyrcana **269**
    × O. tenthredinifera <u>404</u>, **405**
spiralis 93
splendida 332
spruneri 264

f. *cretica* 318
   subsp. *gottfriediana* 272
   subsp. *grigoriana* 264
   var. *reinholdii* 322
stavri 301
straussii 325
   subsp. *antiochiana* 325
   var. *leucotaenia* 325
subfusca,
   subsp. *aspea* 227
   subsp. *blitopertha* 231
   subsp. *cinereophila* 232
subinsectifera subsp. *subinsectifera* 217
sulcata 228
×*sulphurea* 225
×*syracusana* 402
taigetica 257
×*tardans* 224
tarentina 330
tarquinia 244
taurica 265
tenthredinifera **223**, <u>224</u>, 397, 405, 430, 436
   subsp. *guimaraesii* 223
   subsp. *spectabilis* 223
   subsp. *villosa* 223
tetraloniae 300
theophrasti 231
thesei 284
thracica 231
thriptiensis 228
transhyrcana 269
   subsp. *amanensis* 248
   subsp. *morio* 265
   subsp. *sintenisii* 265
tremoris 265
trollii 243
tyrrhena 333
ulupinara 301
ulyssea 223
umbilicata 33, **313**, 390, 431
   subsp. *beerii* 313
   subsp. *bucephala* **314**
   subsp. *calycadniensis* 317
   subsp. *flavomarginata* <u>316</u>, **317**
   subsp. *lapethica* <u>316</u>, **317**, 318, 401
   subsp. *latakiana* **318**, <u>319</u>
   subsp. *latilabris* 317
   subsp. *rhodia* 313
   subsp. *umbilicata* **313**, <u>314</u>, 401
urteae 231
vallesiana 232
vasconica 326
vernixia 218
   subsp. *ciliata* 218

   subsp. *lusitanica* 218
   subsp. *orientalis* 218
vetula 302
×*vicina* 33, **335**, 431
   nm. "calypsus" <u>**335**</u>
   nm. "heterochila" **336**, <u>337</u>
   nm. "holubyana" **336**, <u>337</u>
vitorica 252
zagoriana 244
zeusii 257
zinsmeisteri 336
ziyaretiana 288
zonata 228
Orchidium arcticum 85
ORCHIS 2, <u>5</u>, 7, 8, 10, 14, 25, 27, 31, 33, **177**, 362, 420, 422
   abortiva 79
   adenocheila 181, <u>**182**</u>, 183, 428
   albanica 384
   anatolica 13, 14, 191, <u>**192**</u>, <u>193</u>, 195, 429
      subsp. *sitiaca* 191
      subsp. *troodi* 192
   ×*angusticruris* 409
   angustifolia var. *lapponica* 160
   anthropophora 33, <u>176</u>, **177**, 406, 420, 428
      × O. *italica* <u>404</u>, **405**
      × O. *militaris* **406**, <u>407</u>
      × O. *purpurea* **406**, <u>407</u>
      × O. *simia* **406**, <u>407</u>
   apifera 240
   ×*bergonii* 406
   ×*beyrichii* 409
   bifolia 100
      var. *latissima* 102
   ×*bivonae* 405
   boryi 380
   brancifortii **196**, <u>197</u>, 429
   brevicornis 191
   canariensis 191
   caprina 345
   caucasica 183
   cazorlensis 188
   champagneuxii 385
   chlorantha 105
   chlorotica 368
   collina 368
   comperiana 339
   conica 213
   conopsea 113
      var. *densiflora* 110
   cordigera 155
   coriophora 372
      subsp. *fragrans* 372
      subsp. *sancta* 371

INDEX OF SCIENTIFIC NAMES

var. *fragrans* 372
var. *sancta* 371
cruenta 140
cucullata 96
cyrenaica 375
dinsmorei 365
diphylla 95
dulukae 368
elegans 366
  var. *dinsmorei* 365
euxina 138
expansa 379
fedtschenkoi 368
foliosa 147
formosa 342
fragrans 372
fuchsii 148
fuciflora 283
galilaea **178**, <u>179</u>, 422, 428
globosa 174
holosericea 283
×*hybrida* 409
hyperborea 99
iberica 129
ichnusae 207
incarnata 138
  var. *integrata* 156
  var. *ochroleuca* 142
insectifera 214
  subsp. *insectifera* **214**, <u>215</u>
insularis 137
israelitica 380
italica <u>12</u>, **178**, <u>179</u>, 422, 428
kelleri 191
Key to the species of Orchis **428**
lactea 211
  var. *conica* 213
laeta <u>198</u>, **199**, 429
langei 204
latifolia,
  subsp. *baltica* 152
  subsp. *elatior* 156
  var. *coccinea* 145
  var. *dunensis* 160
laxiflora 362
  subsp. *dielsiana* 365
  subsp. *dinsmorei* 365
  subsp. *palustris* 366
  var. *dinsmorei* 365
  var. *palustris* 366
leucoglossa 368
loeselii 83
longicornu 387
  var. *picta* 388

longicruris 178
×*loreziana* 406
×*macra* 406
maculata 147
majalis 152
  subsp. *traunsteinerioides* 172
  var. *occidentalis* 164
mascula <u>3</u>, 6, 8, **203**, 406, 409, 429
  subsp. ichnusae <u>206</u>, **207**
  subsp. laxifloriformis **204**, <u>205</u>, 429
  subsp. mascula 200, <u>202</u>, **203**
  subsp. scopulorum **204**, <u>205</u>
  subsp. speciosa **204**, <u>205</u>
  var. *olbiensis* 200
  var. *olivetorum* 207
  var. *speciosa* 204
    × O. pallens **406**, <u>407</u>
    × O. provincialis <u>408</u>, **409**
masculolaxiflora 204
maurusia 151
mediterranea subsp. *georgica* 134
militaris 16, 19, 33, 181, <u>184</u>, **185**, 390, 406, 409, 428
  subsp. militaris <u>184</u>, **185**
  subsp. *simia* 181
  subsp. stevenii <u>184</u>, **185**
  var. *adenocheila* 182
  var. *purpurea* 182
    × O. purpurea <u>408</u>, **409**
    × O. simia <u>408</u>, **409**
miniata 122
morio 382
  subsp. *champagneuxii* 385
  subsp. *syriaca* 388
  var. *caucasica* 384
  var. *champagneuxii* 385
  var. *longicornu* 387
  var. *mascula* 203
  var. *provincialis* 199
odoratissima 114
olbiensis **200**, <u>201</u>, 429
orientalis subsp. *cilicica* 142
ovalis 204
pallens 19, **196**, <u>197</u>, 406, 429
palustris 366
  subsp. *laxiflora* 362
  subsp. *pseudolaxiflora* 365
  var. *elegans* 366
  var. *robusta* 368
panormitana 191
papilionacea 376
  subsp. *alibertis* 376
  subsp. *balcanica* 376
  subsp. *heroica* 376

subsp. *palaestina* 379
var. *cyrenaica* 375
var. *grandiflora* 379
var. *messenica* 376
patens **191**, 429
subsp. canariensis <u>190</u>, **191**
subsp. *nitidifolia* 188
subsp. patens <u>190</u>, **191**
subsp. *spitzelii* 186
pauciflora **200**, <u>201</u>, 429
×penzigiana 409
picta 388
subsp. *libani* 388
var. *albanica* 384
var. *caucasica* 384
var. *champagneuxii* 385
pinetorum 203
prisca 188
provincialis 14, <u>198</u>, **199**, 409, 429
subsp. *pauciflora* 200
pseudolaxiflora 365
punctulata 13, <u>180</u>, **181**, 182, 428
subsp. *adenocheila* 182
subsp. *galilaea* 178
subsp. *schelkownikowii* 181
var. *galilaea* 178
var. *sepulchralis* 181
purpurea 19, 33, 181, **182**, 390, 406, 409, 428
subsp. caucasica **183**
subsp. purpurea **182**, <u>183</u>
var. *caucasica* 183
× O. simia <u>408</u>, **409**
purpurella 166
pyramidalis 370
quadripunctata <u>194</u>, **195**, 196, 429
subsp. *anatolica* 192
var. *boryi* 380
robertiana 339
robusta 368
romana 132
var. *guimaraesii* 134
rubra 376
saccata 368
saccifera 151
sambucina 131
sancta 371
scopulorum 204
sesquipedalis 169
×sezikiana <u>195</u>
simia 13, <u>23</u>, 178, <u>180</u>, **181**, 406, 409, 428, 458
subsp. *taubertiana* 181
sitiaca <u>190</u>, **191**, 192, 429
sparsiflora 368
sphaerica 177

sphagnicola 170
spitzelii **186**, 429
subsp. cazorlensis **188**, <u>189</u>
subsp. nitidifolia **188**, <u>189</u>, 191
subsp. spitzelii **186**, <u>187</u>
×spuria 406
stevenii 185
syriaca 388
taubertiana 181
tenera 203
tridactylites 94
tridentata 213
troodi 192
urvilleana 172
ustulata 208
var. *aestivalis* 208
vomeracea 355
subsp. vomeracea <u>354</u>, **355**

P

Paludorchis 362
dinsmorei 365
laxiflora 362
palustris 366
var. *elegans* 366
pseudolaxiflora 365
robusta 368
Peramium macrophyllum 89
PLATANTHERA <u>5</u>, 25, **99**, 421
algeriensis **106**, <u>107</u>, 425
atropatanica 102
azorica **106**, <u>107</u>, 425
bifolia 6, 7, 99, **100**, <u>101</u>, 102, 409, 413, 425
subsp. *chlorantha* 105
subsp. *latiflora* 100
subsp. *osca* 102
var. bifolia **100**, <u>101</u>
var. latissima <u>102</u>, **103**
× Gymnadenia conopsea <u>412</u>, **413**
× P. chlorantha <u>408</u>, **409**
chlorantha 6, 7, 19, 99, 102, <u>104</u>, **105**, 106, 409, 425
subsp. *algeriensis* 106
subsp. *holmboei* 105
fornicata 102
holmboei <u>104</u>, **105**, 106, 425
×hybrida 409
hyperborea <u>98</u>, **99**, 425
Key to the species of Platanthera **425**
kuenkelei 102
var. *sardoa* 102
lesbiaca 105
micrantha <u>108</u>, **109**, 425
muelleri <u>102</u>, **103**, 425

INDEX OF SCIENTIFIC NAMES

oligantha **100**, <u>101</u>, 425
pollostantha <u>108</u>, **109**, 425
PONERORCHIS 25, 27, **96**, 421
  cucullata 1, **96**, <u>97</u>, 421
PSEUDORCHIS 25, **96**, 421
  albida <u>6</u>, **96**, <u>97</u>, 421
  f. *straminea* 99
    subsp. albida **96**, <u>97</u>
    subsp. straminea <u>98</u>, **99**
  frivaldii 110
  straminea 99

S

Satyrium,
  albidum 96
  hircinum 342
  maculatum 207
  nigrum 118
  repens 89
  viride 126
SERAPIAS 2, 10, 25, 33, **348**, 420
  ×ambigua 410
  anthropophora 177
  aphroditae 356
  apulica 356
  athwaghlisia 359
  azorica 348
  bergonii 355
  cilentana 355
  cordigera **348**, <u>349</u>, 350, <u>351</u>, 410, 433
    var. *cretica* 348
    var. *mauritanica* 348
    × S. lingua **410**, <u>411</u>
  cossyrensis 348
  elsae 352
  feldwegiana 356
  gregaria 358
  helleborine 56
  ionica 356
  istriaca 356
  Key to the species of Serapias **433**
  latifolia var. *atrorubens* 51
  levantina 356
    var. *dafnii* 356
  lingua <u>**360**</u>, <u>361</u>, 410, 433
  neglecta **352**, <u>353</u>, 433
  nidus-avis 79
  nurrica <u>**351**</u>, 433
  occidentalis 350
  olbia <u>**358**</u>, 433
  orientalis **356**, <u>357</u>, 433
    var. *carica* 356
    var. *monantha* 356
    var. *sennii* 356

    var. *siciliensis* 356
    var. *spaethiae* 356
  parviflora **352**, <u>353</u>, 356, 433
    subsp. *laxiflora* 355
  patmia 356
  perez-chiscanoi <u>**350**</u>, 433
  politisii 356
  strictiflora 352, <u>**359**</u>, 433
  vomeracea 33, 350, **355**, 356, 410, 433
    subsp. laxiflora <u>354</u>, **355**, 356
    subsp. *orientalis* 356
    var. *guadarramica* 355
Serapiastrum 348
×Serapicamptis rousii 410
SPIRANTHES 25, **90**, 424
  aestivalis **90**, <u>91</u>, 93, 424
  australis 90
  autumnalis f. *parviflora* 93
  cernua,
    × S. odorata 90
  Key to the species of Spiranthes **424**
  lucida 90
  odorata 90
  romanzoffiana 20, **90**, <u>91</u>, 424
  spiralis 19, 90, <u>92</u>, **93**, 424
  stricta 90
STEVENIELLA 2, 25, **336**, 421
  satyrioides 7, **336**, <u>337</u>, 421
Stevenorchis satyrioides 336

T

TRAUNSTEINERA 25, **174**, 419
  globosa 14, **174**, <u>175</u>, 428
    subsp. *sphaerica* 177
  Key to the species of Traunsteinera **428**
  sphaerica <u>176</u>, **177**, 428

V

Vermeulenia 362
  bruhnsiana 376
  collina 368
  cyrenaica 375
  papilionacea 376
    var. *aegaea* 376
    var. *alibertis* 376
    var. *grandiflora* 379
    var. *messenica* 376
    var. *palaestina* 379
    var. *vexillifera* 379

459

Orchis simia
Greece, Peloponnese (25)

INDEX OF COMMON NAMES

A
Adriatic lizard orchid 344
alpine vanilla orchid 117
Anatolian orchid 192
Armenian marsh-orchid 145
Atlantic bee orchid 240
Atlas orchid 191
Austrian vanilla orchid 118
autumn lady's tresses 93

B
bee orchid 240
bird's nest orchid 79
bog fragrant orchid 113
bog orchid 82
Bory's orchid 380
Brancifort's orchid 196
broad-leaved helleborine 56
broad-leaved marsh-orchid 152
bug orchid 372
bumble-bee orchid 221
burnt-tipped orchid 208
Buschmann's vanilla orchid 122

C
calypso 85
Canary Islands' habenaria 94
common spotted-orchid 148
common tongue-orchid 360
Comper's orchid 339
Coralroot orchid 86
Cornel's vanilla orchid 121
creeping lady's tresses 89
Cretan bee orchid 318
Cretan helleborine 45
Crimean marsh-orchid 129
Cyprus bee orchid 322

D
dark-red helleborine 51
dense-flowered fragrant orchid 110
dense-flowered orchid 207
dingy bee orchid 228
dune helleborine 70

E
early marsh-orchid 138

early-purple orchid 203
early spider orchid 244
eastern marsh helleborine 48
eastern tongue-orchid 356
eastern violet helleborine 55
eastern woodcock orchid 313
elder-flowered orchid 131
European bird's nest 1
eyed bee orchid 275

F
fen orchid 83
fly orchid 214
four-spotted orchid 195
fragrant bug orchid 372
fragrant orchid 113
Frivald's orchid 110
frog orchid 126

G
Galician orchid 137
Galilean orchid 178
gennaria 95
ghost orchid 1, 80
giant orchid 339
globe orchid 174
greater butterfly orchid 105
green butterfly orchid 106
green-flowered helleborine 74
green-winged orchid 382
Greuter's helleborine 66

H
heart-lipped tongue-orchid 348
heath spotted-orchid 147
Hochstetter's butterfly orchid 106
Holmboe's butterfly orchid 105
holy orchid 371
horseshoe orchid 272
hybrid tongue-orchid 358

I
Iceland butterfly orchid 99
Irish lady's tresses 90
Israeli green-winged orchid 380

462 | INDEX OF COMMON NAMES

K

Kotschy's helleborine 46
Kurdish helleborine 45

L

lady orchid 182
lady's slipper orchid 38
late spider orchid 283
lax-flowered orchid 362
lax-flowered tongue-orchid 355
lesser butterfly orchid 100
lesser twayblade 76
Libyan butterfly orchid 375
lizard orchid 342
long-lipped tongue-orchid 355

M

Madeiran early-purple orchid 204
Madeiran lady's tresses 89
Madeiran orchid 147
man orchid 177
marsh helleborine 51
military orchid 185
milky orchid 211
mirror of Venus orchid 218
monkey orchid 181
Müller's butterfly orchid 102
Müller's helleborine 73
musk orchid 93

N

naked-man orchid 178
narrow-lipped helleborine 69
neottianthe 96
northern butterfly orchid 100
northern marsh-orchid 166

O

one-leafed bog orchid 85

P

pale-flowered orchid 196
Perez-Chiscano's tongue-orchid 350
pink butterfly orchid 376
Provençal orchid 199
punctate orchid 181
purple lady's slipper 41
pyramidal orchid 370

R

red helleborine 42
red vanilla orchid 122
Reinhold's bee orchid 322

Roman orchid 132
Rune's vanilla orchid 124

S

saccate-lipped orchid 368
Sardinian early-purple orchid 207
Sardinian tongue-orchid 351
sawfly orchid 223
scarce tongue-orchid 352
short-spurred butterfly orchid 109
short-spurred fragrant orchid 114
showy early-purple orchid 204
small white orchid 96
small-flowered butterfly orchid 109
small-flowered tongue-orchid 352
small-leaved helleborine 55
southern marsh-orchid 156
Spanish early-purple orchid 204
Spanish vanilla orchid 114
sparse-flowered orchid 200
Spitzel's orchid 186
spotted lady's slipper 41
spurred helleborine 42
Steven's orchid 336
straight-flowered tongue-orchid 359
Strauss's bee orchid 325
summer lady's tresses 90
sword-leaved helleborine 45

T

Taurus bee orchid 325
Taurus lizard orchid 348
Tenerife giant orchid 341
toothed orchid 213
Trabut's limodore 80
twayblade 76

V

vanilla orchid 118
violet helleborine 65
violet limodore 79

W

wasp orchid 243
white helleborine 46
Widder's vanilla orchid 121
woodcock orchid 301

Y

yellow bee orchid 225
Yugoslavian vanilla orchid 117